建筑是怎样炼成的
HOW THE ARCHITECTURE WAS TEMPERED

先锋空间 | HKASP 编

中国林业出版社

SYNOPSIS

All students studying architecture or urban planning will want to know what the right design method is when they begin to learn design. The situation is basically no different to teachers. Corresponding to the beginners' question, teachers will ask what the right way to teach design is. There are at least two different possibilities in the main dimensions: one is the selected distinguished examples, and the other is about the methods to develop solutions. This book adopts the first one and elaborates specific design process to everyone in the form of selected distinguished examples.

This book is not the practice of architectural theory, but the display of excellent practical achievements under the guidance of design science. Architectural theory mainly shows the historical architects' different positions and styles and their reflections on architecture. The objects of the architectural theory cover the dominant ideas of a certain period or an architect, a design team. Different from this theory, the theory of design science is about the phenomenon of architectural works and similar objects. Its core concern is not on history but on structure. Its questions are: What is design? How is design carried out? What do designers mainly do? What kind of rationality do they follow?

导读

建筑学或城市规划的学生开始学习设计之初，都会希望知道什么才是正确的设计方法。这种情况对老师而言基本上也没有什么不同，与初学者的问题相对应，老师也会问，什么才是正确的方法来进行设计教学。在主要的维度上至少有两种不同的可能性，一个是精选优秀范例，一个是讲如何发展解决方案的方法。而本书采用前一种方法，采用精选优秀范例的形式，跟大家阐述具体的设计过程。

本书不是建筑理论的实践，而是在设计科学的指导下优秀实践成果的的展示。建筑理论主要显示了历史上建筑师的不同立场和风格以及对建筑的反思，其对象涉及到某个时期的主导思想或者一个建筑师、一个设计小组的主导思想。与这种理论不同，设计科学要发展的理论是关于建筑作品和相似对象的现象，它主要关注点不在历史而是结构，它的问题是：什么是设计？设计是怎样进行的？设计者主要做什么？他们遵循什么样的合理性？

ABOUT DESIGN

What is design?

The definition of design should be sufficiently general and has covered a variety of different designs including industrial design, landscape design, architectural and urban design. There are many different definitions of design in literature. They individually emphasise different aspects. Some focus on the visual aspect, some put particular emphasis on the problem-solving process, and some concern on information processing. But all definitions admit such a fact, that is, design is a behaviour, not a result and the definition of design has nothing to do with that of "architecture".

Design is :
-A process of developing reasonable solutions.
-A process of developing knowledge about problems related to solutions. These problems mix with emerging visual imaginations at the same time and become the objects for further refinement and development.
-A behaviour of processing purposes, conditions, means and other information in the interaction process of creating and reducing possible plans.
-A behaviour aiming at putting a plan into practice and getting the desired results caused by the realisation of the plan.

关于设计

设计是什么?

设计的定义应该足够普遍已涵盖各种不同的设计,包括工业设计、景观设计、建筑和城市设计,文献中对设计有许多不同的定义,它们各自强调不同的方面。有些主要关注视觉方面,有些侧重问题解决的程序,有些侧重信息处理。但所有的定义都承认这样一个事实,设计是一个行为,不是一个结果,也跟"建筑"的定义无关。

设计是

一个创造合理解决方案的过程。

一个开发关于解决方案相关问题的知识的过程,这些问题同时与出现的视觉想象相混合,成为进一步推敲和发展的对象。

是一个在创造和缩减可能方案多样性的交互过程中处理关于目的、条件、手段等信息的行为。

是一个旨在将计划付诸实施的行为,并且如果计划实现将会导致期望的结果。

ABOUT DESIGN PROCESS

Design is complex and even a very simple problem will be very complex in design. Firstly, it is difficult to define the scope of problems and determine the space to find the answers. Secondly, it can be seen that the solutions of problems depend on the definitions of relevant aspects which influence the definitions of the problems in turn. The final decisions do not come from particular rational routes. They are the results of bargaining among architects, business owners, market analysts, manufacturers, economists, marketing experts and other stakeholders.

Dealing with complex problems means it is necessary to face conflicting value systems of a variety of parties and face the difficulty of limiting or reducing solutions, because the design solutions are unlimited theoretically and a large number of factors from all aspects and their unusual interaction ways need to be taken into account.

Among the many factors, there are three aspects having a relatively large impact on design: one is contexts, one is human factors, and another one is functions. Most designers will begin to design from the three aspects. And according to specific projects, every designer will have a certain emphasis or priority when considering these three aspects.

关于设计过程

设计是复杂的,在设计中即使非常简单的问题也会非常复杂。首先是难以界定问题的范围并确定寻找答案的空间,其次,我们可以看到问题的解决方案取决于相关方面的定义,而这些方面又影响着对问题本身的定义。最后的决定并非来自于某一特定的理性路线,而是建筑师、企业主、市场分析师、生产商、经济师、营销专家和其他利益相关者之间讨价还价的结果。

处理复杂性问题意味着要面对各种参与者之间冲突的价值体系、面对限定缩小解决方案范围的难度,因为理论上设计的解决方案是无限的,需要考虑大量方方面面的因素以及它们不同寻常的相互作用方式。

在众多因素里,主要有以下三个方面对设计有比较大的影响,一个是文脉,一个是人的因素,还有一个是功能,设计师们大多在做设计时都从这三个方面着手。而根据具体的项目,每个设计师在考虑这三个方面时,都会有一定的侧重点,或者优先级。

ABOUT CONTEXTS

No houses are isolated and they are connected with the outside world, linked with law, society, environment, profession, economy, politics, culture, etc. When facing these complex information, architects need to be dedicated and treat the systems including the design objects and their site conditions as an entity. Architectural design is an intervention to the real world. When the works of architects meet with the real world, there is causality between them. Context is a part of the real world and it exists before a new building is created. From the perspective of context treatment, it can be seen that architecture is neither autistic, specific behaviour, nor an artist's autonomous aesthetic statement. However, as an intervention to the real world, architectural design is often a behaviour of gaining inspiration through contexts. Context treatment may be seen as a constraint, but architects still have free space to treat contexts.

Contexts are not independent objects or ideas like geometric graphics, colours, etc. Contexts are always related to other things which are the main concerns and are implanted into the contexts. Context indicates a relationship. Architects must take contexts into account when they begin to design. Sometimes, context is an intervention, a basis, a motivation for design.

Context can be defined as a system composed of all elements in a particular piece of the real world the architect intends to intervene and consider. These "elements" have different natures. They can be physical elements like adjacent buildings, can also be spiritual elements like happiness and satisfaction and so on, but also be mental elements such as knowledge, habits, customs and other things about a culture. By extension, they can also describe the considerations affecting ecosystems from many aspects. To this extent, context has represented the whole known network of all possibilities of influence factors in a general sense, and is the source of stimulating design inspiration and the results of

关于文脉

没有一座房子是孤立的，它们是与外界相连的，与法律、社会、环境、职业、经济、政治、文化等相联系，建筑师们在面对这些复杂性的信息时，需要专注的将包含设计对象及其所处场地情况的系统作为一个整体来处理。建筑设计是对现实世界的干预，建筑师的作品与现实世界相遇，彼此并非没有因果，文脉就是这个现实世界的一部分，在一个新建筑加入之前就已经存在。从处理文脉的角度能够看出，建筑既不是自闭的、特异的举动，也不是艺术家的自主性美学陈述，然而，作为一种对真实世界的干预，建筑设计往往是一种通过文脉来赢得其灵感的行为。对文脉的处理可能会被看作是一种制约，但建筑师还是有处理文脉的自由空间。

文脉不是一个像几何形、色彩等独立存在的事物或构想，文脉总是与其他的一些事物相关，这些事物是主要关注的对象并且植入于文脉之中，文脉指明了一种关系。文脉是建筑师开始设计时必须考虑到的问题，有时，文脉对于设计是一种干预，一种依据，一种激发。

文脉可以定义为建筑师打算干预和考虑的某个现实世界片段中所有元素集合而成的系统。这些"元素"具有不同的本性，它们可以是物质元素如相邻建筑物，也可以是精神元素如幸福和满足等，同样也可以是心理元素如关于一种文化的知识、习惯、习俗等，扩展开来还有描述影响生态系统诸多方面的考虑等等。从这个程度上讲，文脉从普遍意义上代表了影响因素所有可能性的整个已知

these design interventions in various systems such as villages, cities, regions, etc.

Contexts contain many elements, such as climate, energy, topography, natural environment, local culture, advance technology, etc. However, among the various aspects of contexts, there is a trade-off in architects' brains or based on the actual project needs. There is no fixed importance degree for the context's various aspects. If an architect considers the artistic side, he will provide a sculptural plan or a conceptual statement; If considering technology, he will display the most innovative materials or structures; If the architect considers ecological guidance, then energy efficiency technology will be adopted and displayed with the architectural form or organisation. Some architects will consider the physical characteristics of the surrounding environment, such as materials, colours, landscapes, etc., while others may consider the symbol needs of developing countries or in cultural process. The cases on the left are for your reference.

Context is not static. The analysing of contexts is also an objective behaviour basically, but context treatment and the responses to the existing contexts are all related to the architects' values, preferences and interests. Besides, the establishment of context consciousness will become the decisive incentives of creativity and judgment in design.

Dimension of freedom of our own imagination must be built in the form of context establishment and treatment. It seems that the value of freedom is freedom itself, but freedom makes action harder: creative concepts' methods, inventions and personal judgments cannot be obtained by relying on context unless there is a more conscious basis.

网络，是我们考虑作为激发设计灵感的源泉以及这些设计干预在各个系统如村庄、城市、区域等的影响结果。

文脉包含的元素众多，如气候，能源，地形，自然环境，当地文化，先进技术等等，但是在文脉的各个不同方面之间，建筑师在大脑中或者因为实际项目的需求，都会有一个权衡。文脉各个方面的重要性没有固定的等级之分，如果建筑师考虑的是艺术方面，他将提供一个雕塑感强的方案或概念性陈述；如果考虑的是技术，他会展示最创新的材料或是结构；如果是生态导向，那么会采用能效技术并将其展示与建筑的形态或建筑的组织。有些建筑师会考虑到周围环境的物理特性如材料、色彩、景观等，而另一些建筑师可能会考虑到发展中国家或文化进程中的对符号的需要。如参左侧考案例所述。

文脉不是一成不变的，对于文脉的分析大多也是客观的行为，但是对于文脉的处理，以及如何对已有的文脉做出回应，都与建筑师的价值观、喜好和兴趣的导向相联系，同时文脉意识的建立将成为设计中创意和判断力的决定性诱因。

在文脉的建立与文脉处理的形式上要建立我们自己想象的自由维度，自由似乎是其本身的价值所在，但它却使行动更为困难：即使依赖于文脉，仍然无法得到创意概念的方法、发明和个人判断，除非建立在更有意识的基础上。

ABOUT PEOPLE

Architecture is not only a single participant in which architects advocate aesthetic concepts or products of developers who are the representatives of functionalism and economic pragmatism, but also a question of how to deal with social needs and the needs of people as building users.

The people, in the entire construction activities, include architects, planners, customers or developers, users and other relevant participants. But users are usually unknown, and because users and architects are two kinds of people with two backgrounds and preference systems, different educational backgrounds, social reference standards and professional environments, they have different understandings to what is success, what is happiness, what is beautiful, what is the full range of daily life, etc.

Therefore, architects need to communicate with users first and develop theoretical background material about the interested themes. Besides, architects have to make a reasonable estimate of the difference of the preferences of architects, users and other relevant people.

Dealing with the questions of users and other participants is a basic problem in actual design process. The work of architects and urban planners is in a social context. Careful investigation of needs, desires and preferences of social actors including users is an important part of our work. Although it always makes design and planning process longer and more complex, the results will be better accepted and will be more sustainable and can enrich our knowledge on the basis of better planning and design.

关于人

建筑不仅是单个参与者，如作为美学概念倡导者的建筑师或作为功能和经济实用主义代理者的开发商的产品，同时也是一个如何处理社会需求以及作为建筑使用者的人的需求问题。
人，在整个建造活动中，包括建筑师，规划师，客户或开发商和使用者等相关参与者。但是通常使用者都是未知的，使用者与建筑师，两个人，两种背景，两种偏好系统，不同的教育背景，不同的社会参考标准，不同的专业环境，对什么是成功，什么是幸福，什么是美观和什么是充分的日常生活情况等等都有不同的理解。
所以，建筑师首先需要与使用者进行沟通，而且应该就关心的主题制定有关这一问题的理论背景材料。同时，建筑师必须对自己的偏好与使用者及其他相关人群偏好的差异做出一个合理的估计。
处理使用者和其他参与者的问题，是实际设计过程中一个基本问题。建筑师和城市规划师的工作处于一个社会文脉之中，包括使用者在内的社会参与者的导向，对需求、愿望和偏好的细致调查是我们工作的重要组成部分，虽然它总是使设计和规划过程更长、更复杂，但结果会更好地被接受和更可持续发展，可以少犯错误，并在更好的规划和设计基础上丰富我们的知识。

ABOUT FUNCTIONS

A function is a collection of activities that serve a particular purpose. It indicates a use of things. The collection can be microscopic and macroscopic according to interests and different concerns.

Before a building becomes a reality, activities that may or should occur within the building will be set generally. Then regions, space, rooms and their properties will be created. Finally, we can develop a list of rooms. In many cases, such as in most architectural design competitions, the lists of rooms are given, but more and more architects and planners still have to develop the lists by themselves or at least give suggestions.

When designing the Blizzard Headquarters, on the basis of determining the original building structure, REX made a detailed analysis of the features of functional units, listed a variety of functions and scenes of activities by taking the function flow as orientation. Then REX set different space according to a variety of scenes. Finally REX connected these units or rooms together in accordance with certain logic.

"Form follows function." The slogan of Sullivan represents an architectural design direction based on functions in modern movement. "Follow" can be interpreted as providing a process emerging from logic.

Function-oriented design ideas can be roughly divided into two directions: one is the interpretation of structuralism which holds that the external forms should obey basic structural rules, bearing and compression functions, force transmission, the separation of bearing structure and envelop enclosure, etc. This idea was the mainstream ideological trend until 1980s and Ludwig Mies Van der Rohe and Walter Gropius are the representatives of this ideological trend.

The other tendency is to carefully analyse the characteristics of functional units and holds that the creation of forms is related to the requirements of various activities and takes the function flow as orientation. The above-mentioned Blizzard Headquarters designed by REX is an example.

关于功能

功能是一个服务于特殊目的的活动的集合，它指明了事物的一种用途，依据对利益和考量的不同关注点，这个集合可以是微观的，也可以是宏观的。

在建筑物成为现实存在之前，一般都会对建筑内可能会发生的活动，应该会发生的活动进行一个设定，然后根据这样的活动及其需求建立区域，空间，房间及其属性，最后我们可以发展出一个房间的列表，很多情况下，如在大多数建筑设计竞赛中，这是给定的，但是越来越多的建筑师和规划师还是得自己进行或者至少要给出建议。

REX 设计的 Blizzard Headquarters，就是在原建筑结构确定的基础上，细致分析了功能单元的特征，以功能流程为导向，罗列了多种功能和活动场景，然后根据各类活动场景设定不同的空间，最后按照一定的逻辑将这些单元或房间联系起来。

"形式追随功能"沙利文的这句口号代表了现代运动中以功能为基础的一种建筑设计导向。"追随"可以被解释为提供由逻辑应运而生的过程。

功能导向的设计思路大概可以分为两个方向：一个是结构主义的诠释，认为外部形式要服从基本的结构法则、承载和受压的功能、力的传递、承重结构和围护结构的分离等等，这种主张进而成为主流思潮，以密斯·凡德罗和沃尔特·格罗皮乌斯为代表人物这一思潮一直持续到 1980 年代。

另一个倾向则是细致分析功能单元的特征，认为形式的创造与各种活动的要求有关，是以功能流程为导向的。上面提到的 REX 设计的 Blizzard Headqurters 就是一个例子。

CONCLUSION

Design intends to change the real world, not to discover or analyse it, or to keep it running or explain it. As a special human activity, design has characteristics different from other activities such as scientific research, organisation, communication and can be supported with the methods adapt to its specific problems. Design problems are complex and show network structure. It is hard to define them because of the unusual interactions of their elements and local schemes, the uncertainty of the effects of the considered schemes, the reliance on judgments, etc. Besides, the processing of knowledge in design process is also special. Except the factual knowledge and instrumental knowledge, moral knowledge, conceptual knowledge and interpretive knowledge are also used. Capacities such as problem solving, analysing, social negotiation, conceptual thinking, future simulation, judging and so on are required for design activities. The capacities can be supported and enhanced with the methods in this book.

It is helpful to clarify complex problems by determining key words or key problems, discussing them according to argumentation rules and using orderly procedures. Every design is an intervention to the existing contexts. The contextual analysis must be based on different aspects not just material aspect. When dealing with questions of relevant people, relevant methods such as empiricism investigation, participation, round tables, mediation and intervention can be adopted. Charts help ttraighten out function problems. Some techniques such as brainstorming or morphological tree diagrams, morphological block diagrams help support originality.

When designers are concerned about the future acts, the challenge will be how to prove the validity of these results through refutation and argument. Any improvement, reinforcement and additional methodology is very important. Our world is too fragile, too sensitive, and too beautiful to be a designed object without carefulness, thoughtfulness and careful consideration.

结语

设计意在改变真实世界，而非发现或分析它，或让它保持运转或解释它，设计作为一项特殊的人类活动具有与其他活动诸如科学研究、组织、沟通等不同的特点，可以由适合其特定问题的方法来支持。设计问题是复杂的并呈网状结构，很难界定它们，其元素和各局部方案的不同寻常的交互关系，所考虑方案影响的不确定性，依赖于判断等等。同时设计过程中知识的加工也是特殊的，除了事实性知识和工具性知识还使用到道义知识、概念性知识和解释性知识。设计活动需要诸如问题解决、分析、社会谈判、概念性思考、模拟未来以及判断等能力，这样的能力可以用本书的方法支持和增强。

通过建立主要关键词或关键问题、根据论证规则讨论它们、使用秩序化的程序等有助于厘清复杂问题。每一个设计都是一个对现有文脉的干预，文脉的分析必须根据不同的方面而不仅仅是物质方面。在处理相关人的问题时，相关的方法如经验主义调查、参与、圆桌会议、仲裁调停等。图表有助于理顺功能问题。一些技巧如头脑风暴或形态树形图、形态框图等有助于支持创意。

设计师关注未来的行为，其挑战将会是如何通过反驳和辩解来证明这些结果的有效性，任何改进、任何强化、任何额外的方法论帮助都是十分重要的，我们的世界太脆弱、太敏感、太美丽，不应该是粗心、考虑不周、没有经过深思熟虑的设计对象。

CONTENTS 目录

014　AXEL SPRINGER CLOUD
Axel Springer Cloud

020　OSLO GOVERNMENTAL QUARTER
奥斯陆政府办公大楼

028　VIRTUAL ENGINEERING IN THE NEW INVENTIVE ECONOMY
创新经济中的虚拟工程中心

042　AXEL SPRINGER MEDIA CAMPUS
Axel Springer 多媒体校园

048　2# WORLD TRADE CENTRE
二号世贸中心大楼

056　REFORMA TOWERS
改革大厦

062　RIVM & CBG HEADQUARTERS BUILDING
RIVM & CBG 总部大楼

070　MIRAI HOUSE, LEIDEN BIO SCIENCE PARK, THE NETHERLANDS
荷兰莱顿生物科学园 Mirai House

076　BLIZZARD HEADQUARTERS
BLIZZARD 办公总部

084　BATTERSEA POWER STATION MALAYSIA SQUARE
巴特西电站马来西亚广场

090　BANKMED HEADQUARTERS II
BankMed 总部 II

096　ETHIOPIAN INSURANCE CORPORATION
埃塞俄比亚保险公司

104　100PP
100pp

110 TAIYUAN TOWER
太原中铁三局科技大厦

116 WOODEN HIGHRISE APARTMENTS
RIVM & CBG 总部大楼

120 NEW KEELUNG HARBOUR AND SERVICE
中国台湾新基隆港服务大楼

126 BALTIC PARNU
Baltic Parnu

132 OCT TOWER
华侨城大厦

142 LAVENUE CROWN
Lavenue crown

OFFICE 办公

CONTENTS 目录

150 Museum of Contemporary Art & Planning Exhibition, Shenzhen
深圳市当代艺术馆与城市规划展览馆

158 Confluence Museum in Lyon, France
法国里昂汇流博物馆

170 HELSINKI CENTRAL LIBRARY
赫尔辛基中央图书馆

178 LEARNING CENTRE POLYTECHNIQUE
巴黎 - 萨克雷大学内巴黎综合理工学校新学习中心

186 DAEGU GOSAN PUBLIC LIBRARY
韩国大邱高山郡公共图书馆

194 HOUSE OF MUSIC, AALBORG, DENMARK
丹麦奥尔堡音乐厅

200 MASSAR CHILDREN'S DISCOVERY CENTRE
MASSAR 儿童探知中心馆

210 NEW TAIPEI MUSEUM OF ART, CHINA
中国新台北市艺术博物馆

220 NEW SCHOOL OF ARCHITECTURE, ROYAL INSTITUTE OF TECHNOLOGY (KTH)
新建筑学院,皇家技术学院(KTH)

228 WUZHONG MUSEUM
吴中区博物馆

234 CITÉ DU CORPS HUMAIN
法国人体博物馆

CULTURE
文化

BUSINESS 商业

244 **BEIJING RIVER CREATIVE ZONE**
北京妫河·建筑创意园区

252 **XINGLONG VISITOR CENTRE**
兴隆访客中心

262 **HSINCHU STONE VILLAGE**
竹岩溪村

270 **KINMEN PASSENGER SERVICE CENTRE**
金门客运服务中心

RESIDENTIAL 居住

282 **1000 MUSEUM**
1000 Museum

288 **PERURI 88**
Peruri 88

294 **DOMINIQUE PERRAULT THE BLADE**
Dominique perrault the blade

300 **LABITZKE-MITTE**
LABITZKE -MITTE

308 **PENTOMONIUM TOWERS**
Pentomonium 双子塔

312 **KINGS HOUSE APARTMENTS**
国王公寓

PROBLEM / 问题

1 SPACE DEMANDS FOR HISTORY AND FUTURE
2 CREATIVE WORK PLACE
3 UNIQUENESS OF MEDIA COMPANY HQ

1. 历史与未来的空间诉求
2. 创造性工作场所
3. 媒体公司总部的独特性

Axel Springer Cloud
Axel Springer Cloud

Ole Scheeren Buro-OS
works
Ole Scheeren Buro-OS

Architect: Ole Scheeren Buro-OS
Client:
Location: Berlin, Germany
Area: 60,000 m²
Function: Complex

设计公司：Ole Scheeren Buro-OS
客户：
地点：德国柏林
面积：60,000 平方米
功能：综合体

DESIGN REQUIREMENTS
设 计 要 求

The contest is aimed to creating more room, especially digital, for media companies. So, the work place needs to be designed in the way that satisfies online media demands. It also should be in consistence with the original HQ on the former site of Berlin Wall.

竞赛的任务是为传媒公司创造更多的空间，尤其是数字空间，因此需要设计工作场所，以满足将来的网络媒体需求，同时与位于柏林墙旧址上的原公司总部相互呼应。

① 设计了一个 30 米高的中庭，建筑的内部被打开，面向与之相邻的 Axel Springer 公司原大楼。

② 项目大楼每层都包含一部分有屋顶覆盖的区域作为传统的工作环境，而剩下的露台未覆盖区域将成为一个非正式的平台以及向公司其他部分传播创新成果的场所。

③ 项目大楼地面第一层向整个城市开放，公司的广播室就安排设在这一层，同时还包括各种活动与展览空间以及各种饭店。

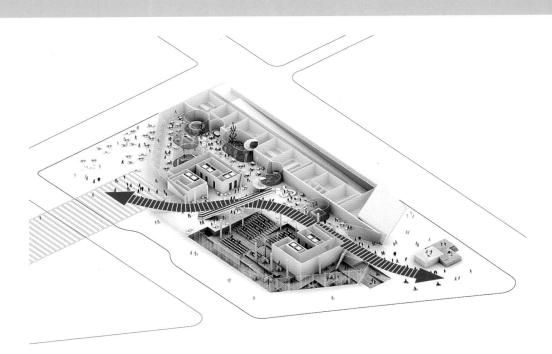

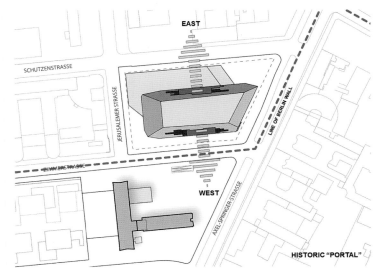

Project Overview

Design concept

Ole Scheeren's design for the 60,000 ㎡ complex in Berlin presents a contemporary counterpart to the historic Axel Springer Tower and an expressive vision for a new collaborative space and identity for the digital media company.

① A 30-metre high atrium is designed to show the building's inside, facing towards the original premise of Axel Springer.

② On each floor, the building comprises certain areas covered with roof as traditional work place, while the rest not roofed as an informal platform and a place for sharing creative work achievements.

③ The first floor of the building is open to the whole city; the company's broadcasting room is seated there, and there are also many activity, exhibition space and restaurants.

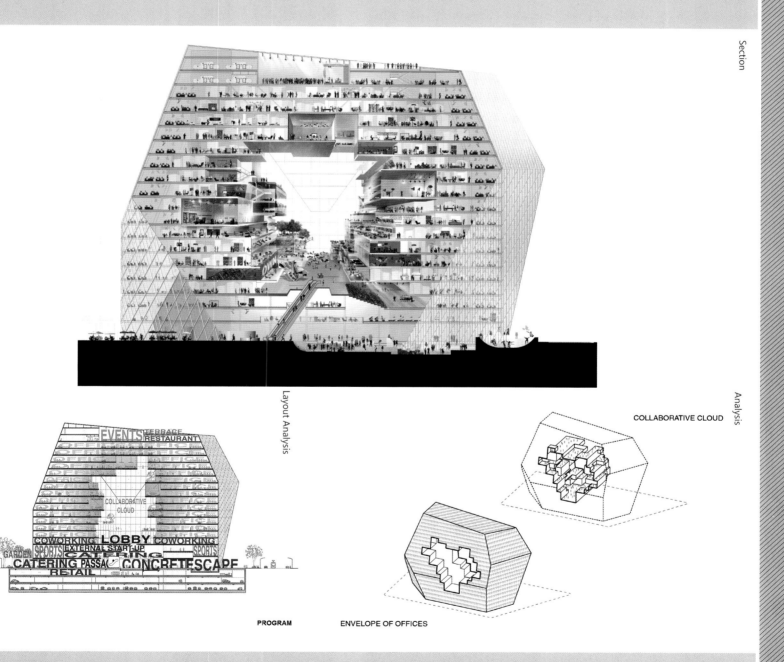

设计理念
奥雷·舍人（Ole Scheeren）为柏林这个 60 000 平方米的综合体设计的方案，以当代手法呼应具有悠久历史的斯普林格大厦，为一个协作空间提供全新想象空间，并将成为一个数字媒体公司的标志。

Location analysis

The architecture reflects the symbolic resonance with the city's urban context. The historically charged site is located along the edge of the former Berlin Wall, which once divided the city and the world, directly opposite the existing Axel Springer Tower, then rising defiantly above the Wall. At the core of the new building floats an urban-scale void, establishing a visual axis between former East and West and conceptually reuniting the two sides. The building emerges as a symbol of transparency and historic awareness.

区位分析

项目基地位于曾经将柏林城和世界一分为二的柏林墙的边缘，与斯普林格大厦正对，建筑将从柏林墙边上升起，对城市文脉作出回应。在新建筑的中央，漂浮着一个巨大的中空地带，视觉上连接起曾经的东西柏林，从概念上重新连接起两个地缘。建筑的通透意象征着透明以及强烈的历史意识。

设计策略

建筑的中空部分被称之为"协作云"，它构成了建筑的核心，并作为新总部的概念和空间的标识。在这里，标识并非是一个具体的物体，而是一个精雕细刻的物理空间，以及空间中所容纳的各种创造、合作和互动。标准化且具灵活性的办公空间沿着建筑刻面设置，身处"云"中，成为整个环境的一部分。

"在一个数字化工作可以在任何地方开展的时代，建筑担负起将人们聚集起来的重要角色，"奥雷·舍人表示，"'协作云'不仅形成了分享交流的实际空间，而且还以其开放形象投射回城市，作为一个共享空间重组企业的多样性。"

建筑创造的多样态城市活动空间和设施便于组合使用，并与城市及周边公共空间互动。大楼的地面层形成一个有着浓郁聚会氛围的空间；一个部分凹陷的混凝土景观提供了多层面的展览、放映和各类活动空间。一条柏林极为典型的公共"通道"横贯建筑，连接起两个周边的公共广场。一个延伸开去的花园覆盖平台层，拥围着整栋大楼；顶层空间则提供特别的活动区域，并有一个室外露台可欣赏到整个柏林城的景色。

Design strategies

A "Collaborative Cloud" forms the heart and nucleus of the building and manifests itself as the conceptual and spatial identity of the new headquarters. Identity is defined not as an object, but as space – a physical void is carved to create flexible permeable places for imagination, collaboration, and interaction. Standardised flexible work spaces are arranged along the perimeter of the facetted building envelope, which dissolve and merge into zones of informal work environments within the Cloud.

"In an age where digital work can be performed anywhere, architecture takes on the critical role of bringing people together," says Ole Scheeren. "The Collaborative Cloud not only forms an actual space of shared ideas and social interaction, but also projects its open image as a powerful gesture towards the city, reuniting a multiplicity of enterprises in a space of shared digital identity."

A diverse array of urban activities and amenities invite inhabitation and interaction with the city and surrounding public domain. The ground levels of the building form a civic base, a marketplace-like environment for gatherings, and a partly sunken concrete-scape offers a multi-level surface for exhibitions, film screenings and ad-hoc urban activities. A public "passage", one of Berlin's ubiquitous typologies, traverses the building and connects the two surrounding public plazas. An extensive garden covers the plinth level and embraces the building, while a rooftop space provides a special area for events, with an outdoor terrace offering views across the city of Berlin.

PROBLEM / 问题

1. **REBUILDING AND EXPANDING**
2. **OPEN AND TRANSPARENT PUBLIC IMAGE**
3. **CONSISTENT WITH GOVERNMENTAL DEVELOPMENT PLAN**

1. 改扩建
2. 公开透明的公众形象
3. 与政府开发计划一致

Oslo Governmental Quarter
奥斯陆政府办公大楼

BIG works
BIG 作品

Architect: BIG
Partners in Charge: Bjarke Ingels, Andreas Klok Pedersen
Collaborators: Hjellnes Consult AS, GULLIK GULLIKSEN AS Landskapsarkitekter, Atelier Ten
Client: Stadsbygg
Location: Oslo, Norway
Area: 124,000 m²
Function: Office building

建筑公司：BIG
合作方负责人：Bjarke Ingels, Andreas Klok Pedersen
合作伙伴：Hjellnes Consult AS, GULLIK GULLIKSEN AS Landskapsarkitekter 和 Atelier Ten
客户：Stadsbygg
面积：124 000 平方米
地点：挪威奥斯陆
功能：办公楼

DESIGN REQUIREMENTS
设计要求

The design of Oslo Governmental Quarter, Norway shall take into consideration the city policies and public engagement.

挪威首都奥斯陆的新的政府中心方案需兼顾城市政策和公众参与。

THE SOLUTION / 解决方式

① 提出一个地形城市公园的方案,设计意图在于为基地内现有历史建筑创造一个新的周边环境。

② 把政府办公大楼创建成一个象征民主,充满生机而且可达性很高的公共空间。

③ 在公共空间内为市民生活的展开留出余地,并同时设置可辐射周边区域的各种服务设施。

 +  +

KVADRATUREN
King Christian IV's *Kvadraturen* neighborhood is a huge part of the DNA of Oslo public space: a simple grid with high density and great diversity between individual blocks.

THE GOVERNMENT QUARTER
The government "neighborhood" on the other hand is a modernist space with large objects on empty squares and layered circulation. It provides symbolic democratic space, but little apparatus for civic life.

CENTRAL PARK
A collective green oasis in a dense urban context.

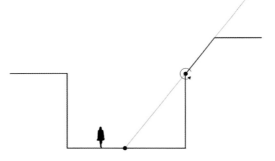

URBAN SET BACK PRINCIPLE
Building shapes are generated from an urban set back and sight line principle. The building height and street width of the immediate surroundings dictate the maximum envelope of the new buildings.

Project Overview / 项目概况

Design concept
BIG proposes a neighbourhood which is integrated as part of the surrounding city, rather than a monumental plan: a democratic public space full of life and easily accessible from all directions. Inspired by the simple, high density grid of Christian IV's Kvadraturen neighbourhood in Oslo, BIG's proposal combines the symbolic democratic space of the existing governmental quarter with the green oasis of New York's Central Park, leaving room for civic life to unfold in the public spaces and amenities that will populate the neighbourhood.

Propose a plan for city park, with the design aimed to creating a new surrounding for the existing historical buildings.	①
Create the governmental building into an energetic and achievable public space that symbolises democracy.	②
Leave room for citizen life in the public space and establish various facilities that serve the areas nearby.	③

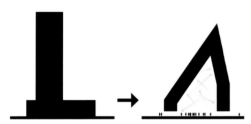

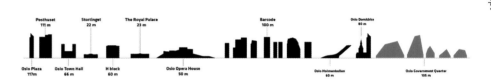

OSLO SKYLINE
The new governmental quarter appears as a mounitan range on the skyline of Oslo. In height, their scale resembles the Oslo Plaza and Posthuset towers, but rendered as a group of crys- taline buildings, oranically growing out of the urban context.
As a symbol, they form a clear and new motive in the skyline: a new mountain range that references the Norwegian land- scape with it's mountains, valleys and fjords.

DEMOCRATIC HIGH RISE
A traditional office highrise seperates the public from staff physically and visually. On the other hand, the democratic highrise invites the public into the ground floor with a generous atrium that allows for visual connection between staff and public.

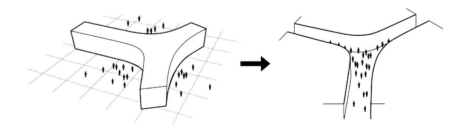

WIND
A vertical facade directs a large part of the wind down towards street level, which can make these areas unpleasent to be in. An inclined facade on the other hand directs less wind down towards the street, improving the quality of the public spaces surrounding the building.

Y-BLOCK
The Y-Block, in the modernist manner, is a sculptural building placed in a large scale open square.

Y-PLAZA
We propose to invert the Y-Block, to turn it into a sculptural urban space rather than a sculptural building in an empty space.

设计理念
BIG 建议把该政府办公大楼创建成一个可融入所在城市的新的周边环境，而不是一个极为庞大的平面：一个象征民主，充满生机而且可达性很高的公共空间。受到奥斯陆 Christian IV's Kvadraturen 周边地区简单的、高密度路网的启发，BIG 在设计方案中，将现有政府办公大楼视作如纽约中央公园的绿地一样的象征民主的空间；设计时，在公共空间内为市民生活的展开留出余地，并同时设置可辐射周边区域的各种服务设施。

A BLANK SLATE
The main challenge is to turn the area into a real neighborhood that can contain the civic life of the city while providing the spatial experience of something larger, shared and collective. In a way, we must combine the qualities of a big shared square (or park) with high density streets full of civic life.

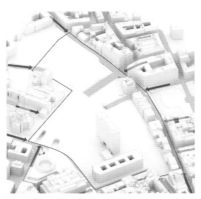

ACCESS POINTS
The surrounding street pattern defines a large number of access points to the area.

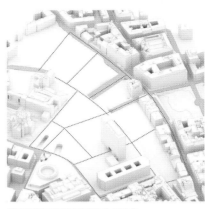

CONTINUATION OF THE URBAN FABRIC
By connecting the access points across the site, through the urban grid, the area becomes an integrated part of the surrounding city.

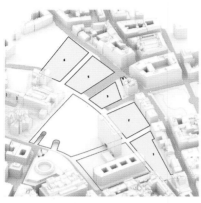

PLOTS
The site is turned into a number of well-proportioned plots that can be developed gradually over the coming decades. The building plots are concentrated along the eastern side of the overall site.

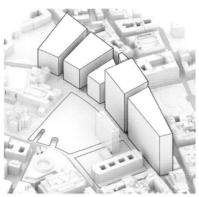

MAXIMUM ENVELOPE
The plots are given a certain maximum height defined by technical restrictions, programmatic demands, etc.

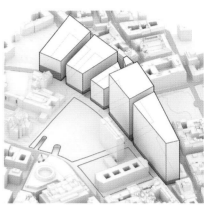

VERTICAL FACADES ALIGNED WITH CONTEXT
Vertical facade heights are defined by the scale of the immediate surroundings.

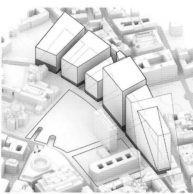

SIGHTLINE SET BACK 1
Each facade is set back according to the sightline from the aligning street level.

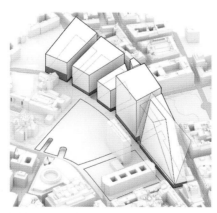

SIGHTLINE SET BACK 2
Building volumes are generated from the immediate context, giving each building a specific form.

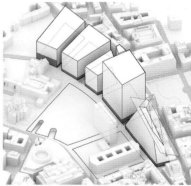

SIGHTLINE SET BACK 4
The result is a height restriction plan specific for each plot. The buildings are organically grown out of their specific context, minimizing over shading and visual impact at street level.

Design strategies

To integrate the existing quarter, surrounding access points are connected across the site, creating a grid with well-proportioned plots that can be developed gradually over the coming decades.

In the West lies a generous park, situating the remaining governmental buildings (the G- and H-Blocks) in a responsive landscape. The iconic Y-Block, which will be torn down as part of the Government's plan, currently occupies a majority of the area where the new park will take root; BIG proposes to create a new kind of urban space in the footprint of the demolished building. The "Y" form is cut into graded topography to create space for restaurants, cafés and public outdoor facilities: Y-Block becomes Y-Plaza.

设计策略

为了将新办公大楼和现有大楼连接起来，周边的各大连接点都在该项目地块上被串联了起来，从而在这块比例得当、错落有致的地块上形成一个路网系统，并且在接下来的几十年中，将逐步得到完善。

在西面，有一个开阔的公园，旧政府办公大楼（G座和H座）就建在这里，与周边景观交相呼应，融于一体。根据政府开发计划的相关要求，这座Y型大楼将被拆除，被其占据的大部分地块将在未来用于建造一个新公园；BIG希望在拆除建筑的痕迹上建立一个新的城市空间。原本的Y形地块被切成不同等级的地形，作为餐馆、咖啡馆和公共设施的户外空间：Y座大楼演变成了Y型广场。

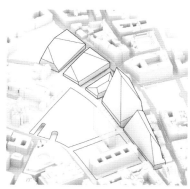  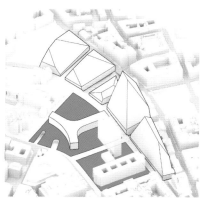 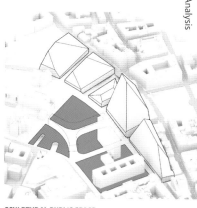

MOUNTAIN RANGE
The crystalline forms are dictated by sight-lines from street level and optimized in relation to solar gain and wind loads, in order to increase efficiency and minimize wind impact at street level.

PARK
A generous park is established on the Western side of the plot, placing the remaining governmental buildings (the G- and H-Blocks) in a new context.

Y-BLOCK
At present, the iconic Y-Block is a monumental modernist building placed in the future park area. What happens when the building is demolished as planned?

SCULPTURAL PUBLIC SPACE
The soft curves of the Y-Block's outline guide the pedestrian flow through the park, dividing the green areas into different zones.

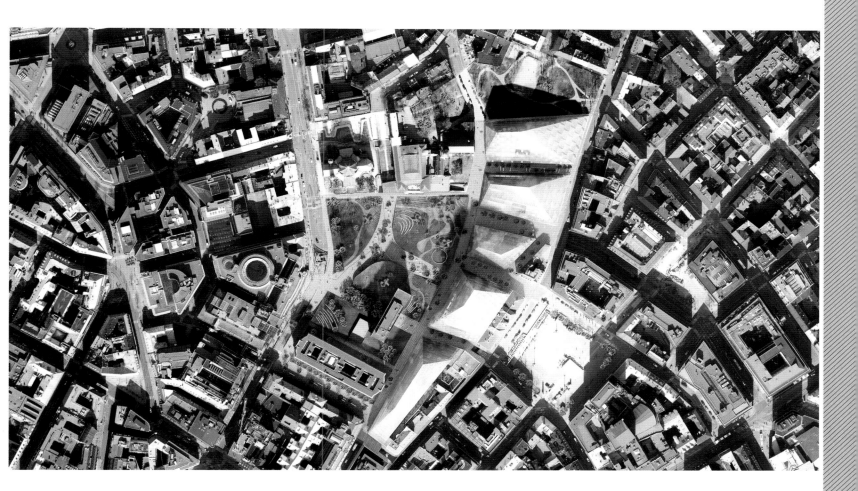

Along the eastern edge of the site, the five buildings in the "mountain range" become democratic high-rises, with workspaces that are visible from public ground floors, also containing generous atria around which collaborative spaces are arranged. Their crystalline forms are generated according to sightlines at street level, optimally designed in relation to solar gain and wind impact for maximum efficiency and quality in the surrounding public areas. The result minimises over-shading while creating unimposing, appropriately scaled space at the neighbourhood-level.

The volumes defined in the "mountain range" become a height restriction plan, a flexible envelope, which over time can become a range of diverse architectures. As symbols, they are organic references to the rich Norwegian landscape: a design driven by Oslo's unique urban context.

沿着该项目地块的东侧看去，有着"山脉"般形态的五栋大楼将成为象征民主的高层建筑物；站在公共首层位置时，可将大楼内各个工作区的情况尽收眼底；同时大楼内部还设有一个开阔的前庭，沿着前庭的周围设置了多个协同合作空间。从站在街道的角度望去，五栋建筑就像五个结晶体一样。采用这种结晶体形式的设计是综合考量当地日照条件和风力影响的结果，以便达到最高的效率并突出周边公共区域的特点。如此一来，在最大限度地减少了建筑的阴影区的同时，为周边环境创造了一个比例得当的、富有亲切感的空间。

有着"山脉"般形态的建筑物将成为一个限制高度的平面布局计划，一个灵活的外壳；随着时间的推移，它们将演变成一系列形式多变的建筑物。作为该地区地标式的建筑物，它们将成为丰富的挪威景观的有机参考物：与奥斯陆独特城市背景融合于一体的设计。

PROBLEM / 问题

① **SUSTAINABILITY**
② **NOVEL OFFICE ENVIRONMENT**
③ **CREATIVE INDUSTRIOUS COMPANY**

1. 建筑的可持续性
2. 新型工作环境
3. 项目为创新性产业公司

Virtual Engineering in the New Inventive Economy
创新经济中的虚拟工程中心

UNStudio
works
UNStudio
作品

Architect: UNStudio
Client: Fraunhofer-Gesellschaft zur Förderung der angewandten Forschung e.V.
Location: Stuttgart, Germany
Gross Floor Area: 5,782 m²
Function: Offices and laboratories

设计公司：UNStudio
客户：弗劳恩霍夫应用研究促进协会
地点：德国斯图加特
总建筑面积：5 782 平方米
功能：办公与实验

DESIGN REQUIREMENTS
设 计 要 求

UNStudio's design for the Centre for Virtual Engineering (ZVE) applies its research into the potential to expand contemporary understandings of new working environments and affect a design approach that creates working environments which stimulate communication, experimentation and creativity through a new type of office building.

UNStudio 在虚拟工程中心（ZVE）的设计中，将其研究应用于挖掘当代对新型工作环境的理解深度及推行关于创造工作环境的设计方案，通过新型的办公建筑推动沟通、实验及创造力。

① 传统的办公室栅格和工作场合的间隔，被结合了大量的视觉连接、偶然或有计划的回忆场所、灵活的办公室实验室及共享的工作地所代替。

② 虚拟工程中心采用图解方式结合了实验室以及研究功能和公共展览区域及参观者的透视路径，加入到一座开放的交际性建筑的概念之中。

③ 建筑通过同时采用紧凑的建筑容积、物质化、灵活的组织、有效的地面应用及大量节能设施的安装，获得德国绿色建筑委员会（DGNB）颁发的金级认证。

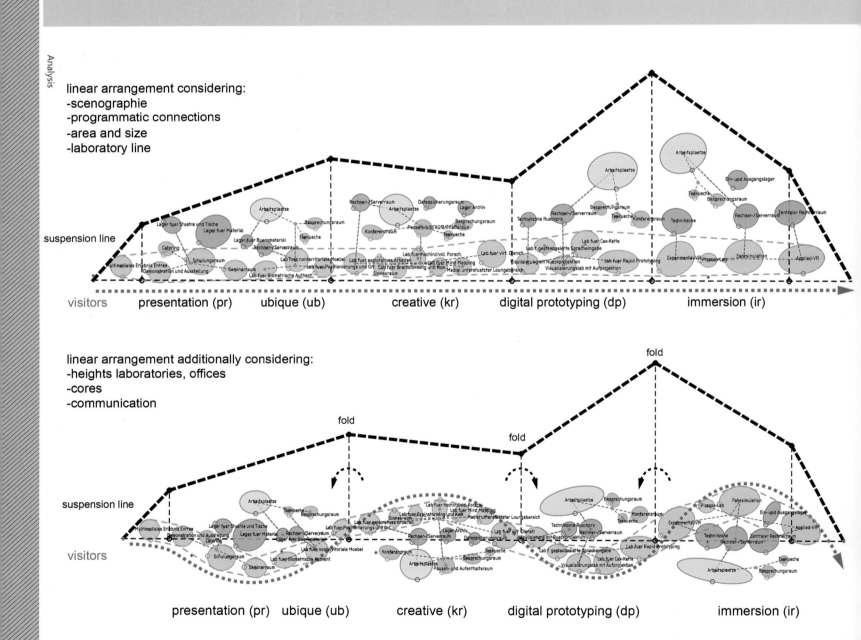

Project Overview / 项目概况

Design strategies

All parts of the ZVE programme are implemented into the spatial organisation of the building. The diagrammatic approach employed combines the laboratory and research functions with public exhibition areas and a scenographic routing of the visitors in an open and communicative building concept.

1. The traditional office grids and work place spacing are replaced by places integrated with lots of visual connection or memories, accidental or planned, flexible office labs and sharing work places.

2. The Centre for Virtual Engineering integrates with labs, research function, public exhibition and visitors through illustration to the conception of an open and interactive building.

3. The building has been granted with Golden Certification by DGNB with its compact building dimension, materialisation, flexible organisation, effective land utilisation and lots of energy-saving facilities.

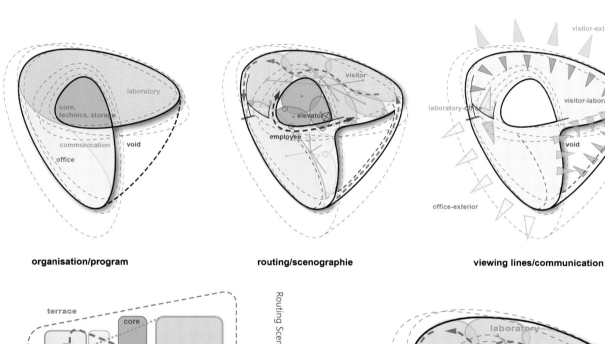

Scheme Organisation

organisation/program   **routing/scenographie**   **viewing lines/communication**

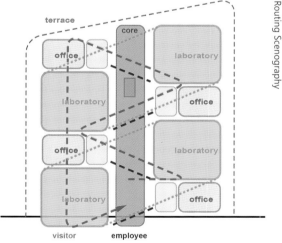

Routing Scenography

routing/scenography

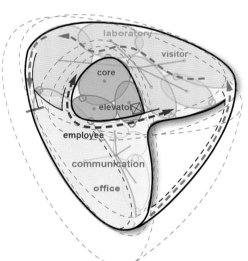

设计策略
虚拟工程中心的各项功能与服务都根据建筑空间布局分布。本着开放、互动的建筑理念，建筑布局采用图解法，将实验室、研究功能与公共展览区、游客参观路线有机地结合起来。

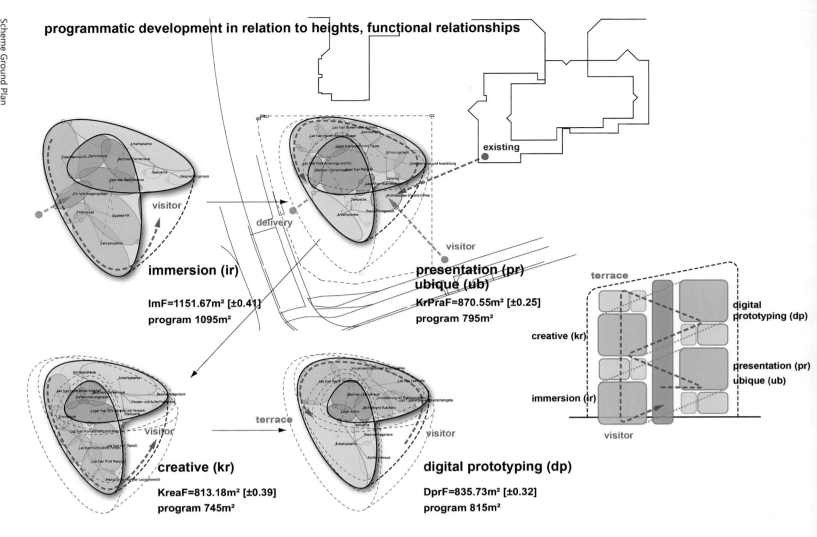

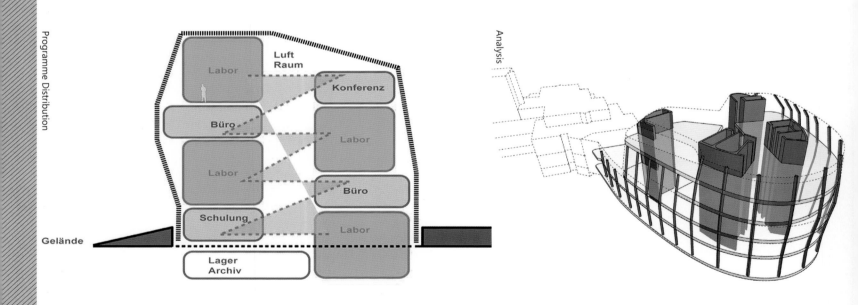

Construction and design elements are integrated within a coherent structure with open and closed elements. The geometry of the floor plan, consisting of curved and straight elements, dissolves into the saw tooth geometry of the facade whilst maintaining the effect of a continuously transforming surface.

建筑构件与设计元素在一个连贯的结构中形成完美的契合，开放与密闭结构并存。在楼层平面图中，曲线与直线元素相互搭接，以自然的形态汇入锯齿形的立面。与此同时，建筑将仍展现不休止的曲面变化效果。

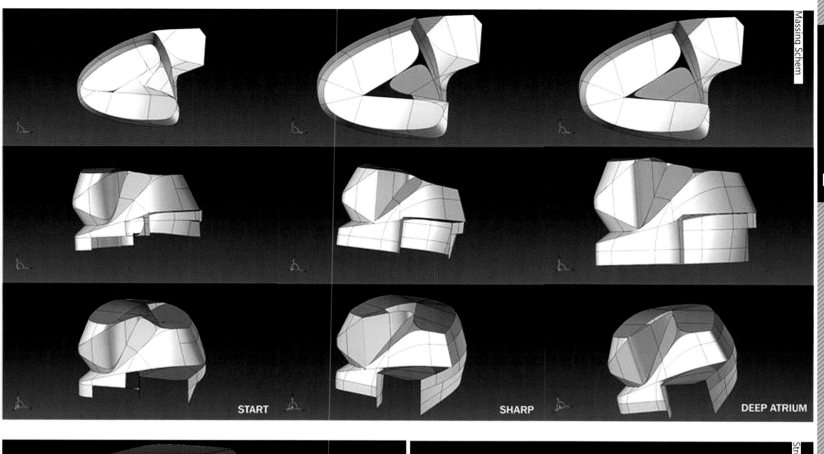

START   SHARP   DEEP ATRIUM

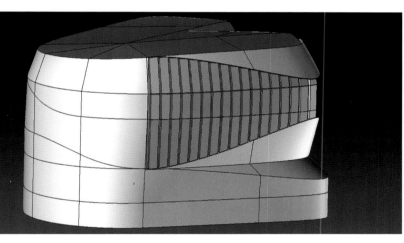
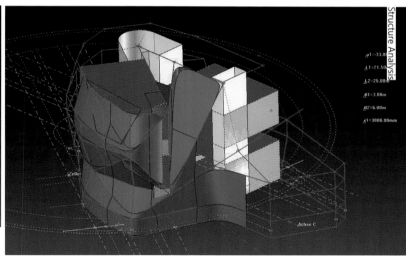
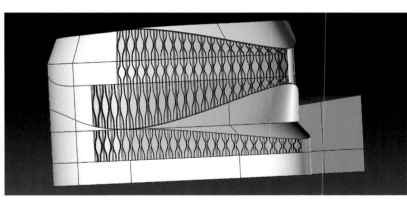
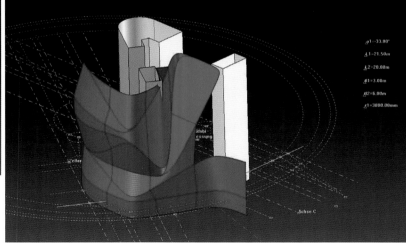

An exceptional level of sustainability was a key consideration from the outset. The building structure partially consists of bubble deck ceilings, providing both an economical alternative to the more commonly used concrete ceilings and reduction in weight, allowing for column-free spaces.

从一开始,如何使其成为绿色节能建筑的标杆成为了我们的一个重点考量对象。建筑构件部分采用塑料球填充的天花板。如此一来,跟常用的混泥土制天花板相比,可以节约更多成本。除此以外,还可减少楼板重量,从而延长结构立柱跨度。

Structure Analysis

sketch "tragwerk" - eingehaengte boxen 13.10.2006

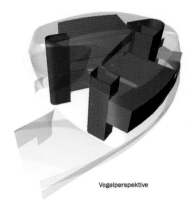

Vogelperspektive

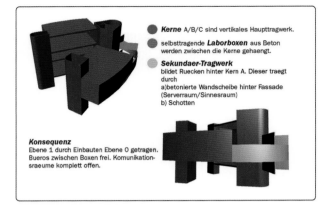

Kerne A/B/C sind vertikales Haupttragwerk.

selbsttragende **Laborboxen** aus Beton werden zwischen die Kerne gehaengt.

Sekundaer-Tragwerk
bildet Ruecken hinter Kern A. Dieser traegt durch
a) betonierte Wandscheibe hinter Fassade (Serverraum/Sinnesraum)
b) Schotten

Konsequenz
Ebene 1 durch Einbauten Ebene 0 getragen. Bueros zwischen Boxen frei. Komunikationsraeume komplett offen.

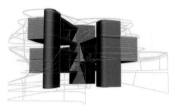
perspektive aus Osten

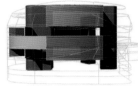
Vogelperspektive aus Sueden

Vogelperspektive

Sketch

Sun Shades Analysis

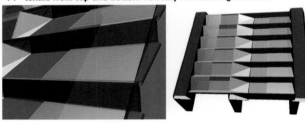

V1 - textile from top and bottom vertically movable in glass level

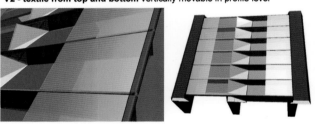

V2 - textile from top and bottom vertically movable in profile level

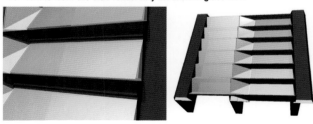

V3 - textile from the side forizontally movable in glass level

Skin

The plot for the Centre for Virtual Engineering ZVE has been used to its maximum in terms of development potential. The rounded shape and optimised building envelope provides a 7% smaller contour than that of a rectangular form of the same area. This also results in a better facade area to volume ratio. The amount of glass facade is only 32%. All spaces along the facade can be ventilated directly by operable window elements.

我们充分挖掘了虚拟工程中心所在地块的开发潜力。跟占有同样用地面积的方形建筑相比，圆柱造型与经过改良的建筑围合使得建筑的周线长度减少7%。同样，如此的造型设计与围合还取得更加理想的立面体积比，从而使玻璃立面

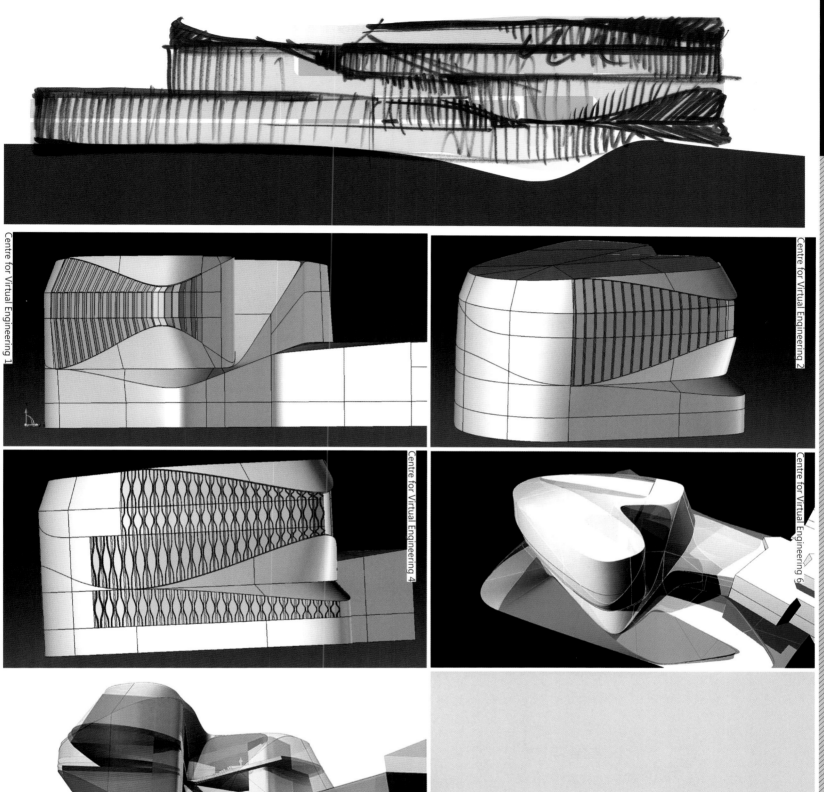

Ceilings without any lintels make it possible for daylight to reflect deep into the spaces, which are additionally supported by daylight lamellas while the sun screens are down. Low maintenance, separable, and recyclable materials have been used for the skeleton as well as for the interior and facade construction.

只占总面积的32%,人们可直接通过调整窗户实现紧挨立面的空间的通风。省去横梁的天花板还可使日光直接折射进入空间深处。此外,当拉下日光屏时,日光薄板还可给空间提供更多的日光。建筑构架、室内构件及立面构件均采用了低维护成本低、可分拆的再生材料。

Facade Band Analysis

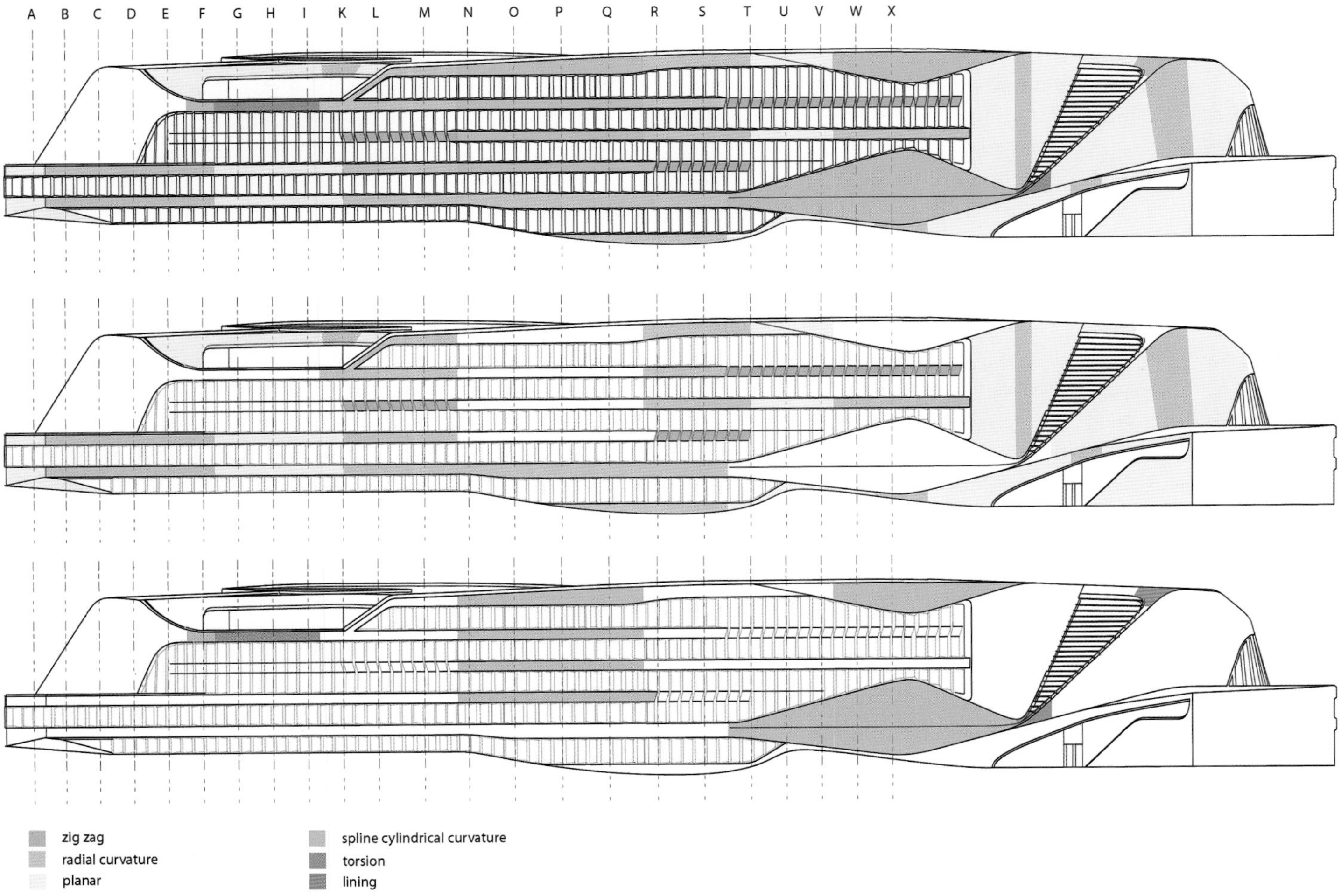

A B C D E F G H I J K L M N O P Q R S T U V W X

- zig zag
- radial curvature
- planar
- spline cylindrical curvature
- torsion
- lining

Window Area Analysis

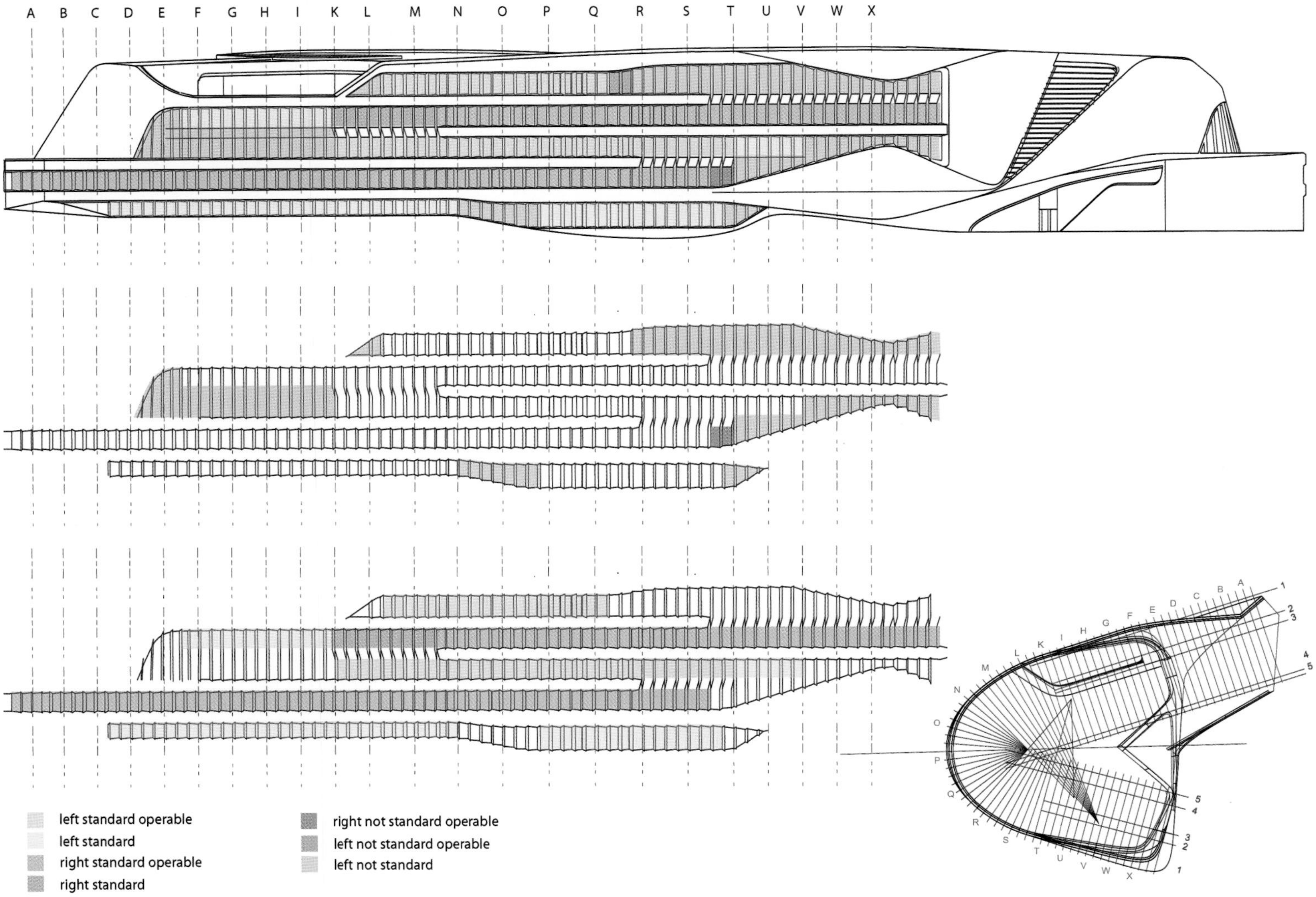

A B C D E F G H I J K L M N O P Q R S T U V W X

- left standard operable
- left standard
- right standard operable
- right standard
- right not standard operable
- left not standard operable
- left not standard

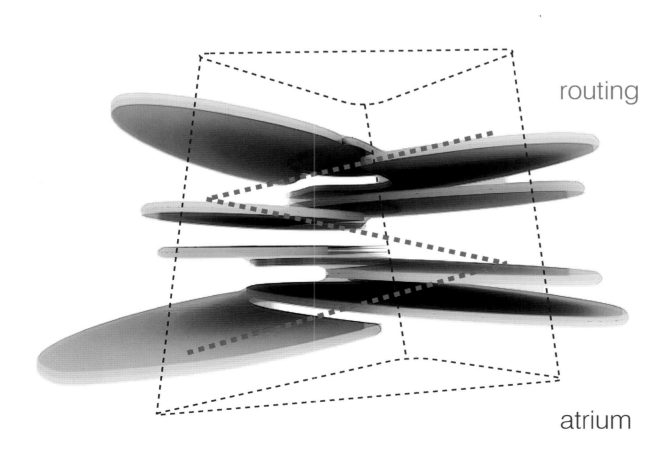

spatial configuration

routing

atrium

Analysis

East Final

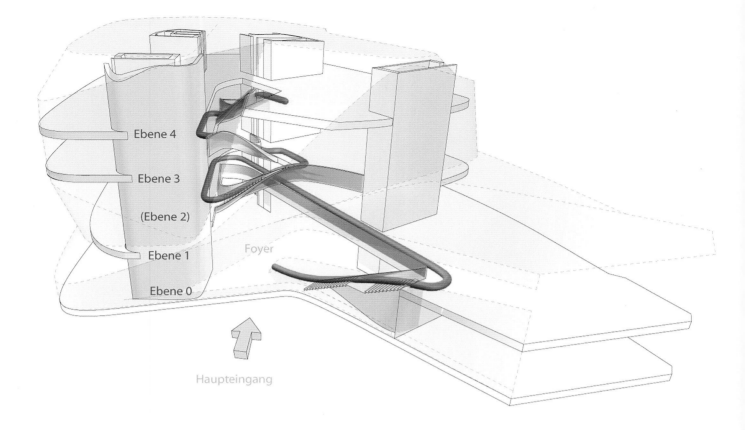

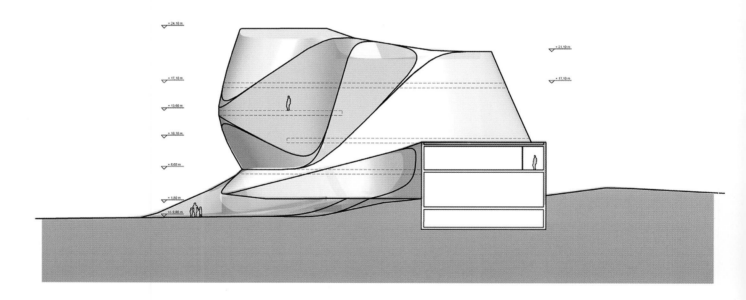

Colour Concept

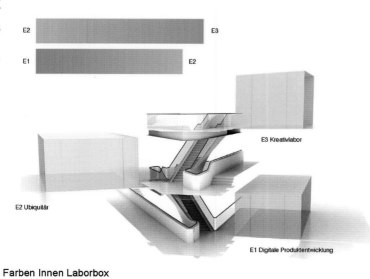

Farben Innen Laborbox

Colour Gradien

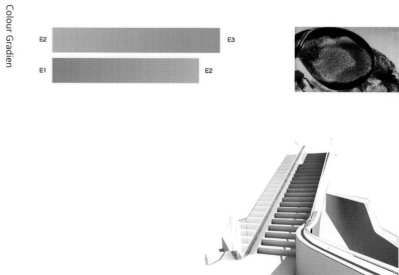

Treppen Farben

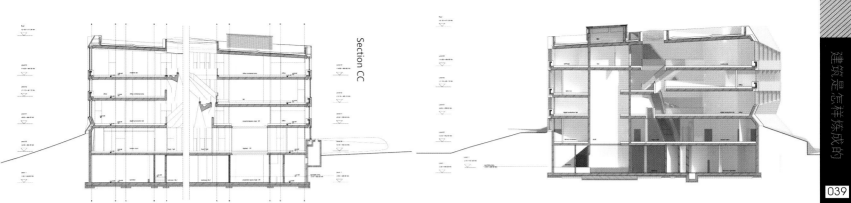

Section CC

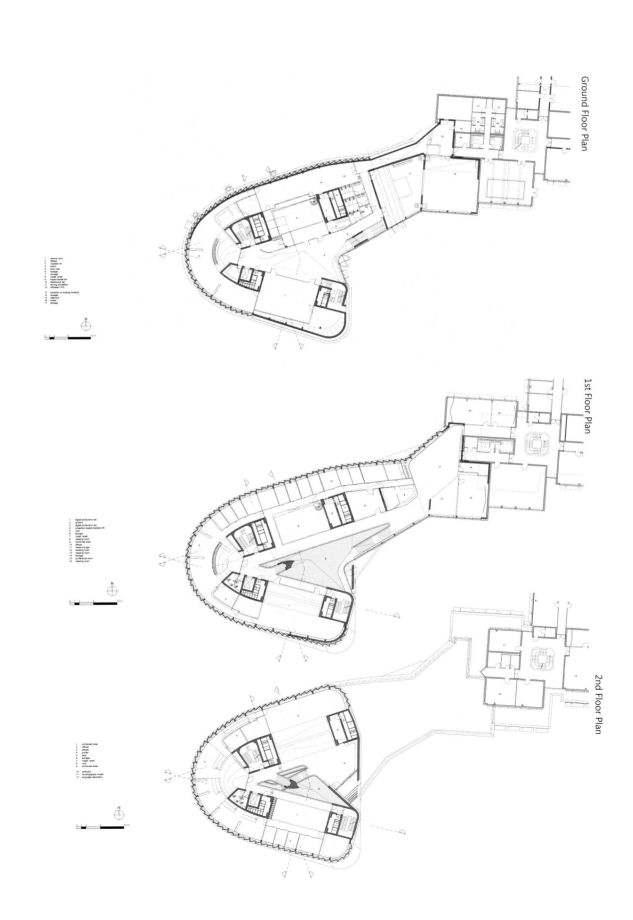

Ground Floor Plan

1st Floor Plan

2nd Floor Plan

PROBLEM / 问题

1. **SPACE DEMANDS FOR HISTORY AND FUTURE**
2. **CREATIVE WORK PLACE**
3. **UNIQUENESS OF MEDIA COMPANY HQ**

1. 历史与未来的空间诉求
2. 创造性工作场所
3. 媒体公司总部的独特性

Axel Springer Media Campus
Axel Springer 多媒体校园

OMA
works

OMA
作品

Architect: OMA
Partners-in-charge: Rem Koolhaas, Ellen van Loon
Project Leaders: Katrin Betschinger, Alain Fouraux and Betty Ng.
Location: Berlin, Germany
Function: Office building

设计公司：OMA
合作方负责人：Rem Koolhaas, Ellen van Loon
项目负责人：Katrin Betschinger, Alain Fouraux and Betty Ng.
客户：
地点：德国柏林
面积：
功能：办公大楼

① 设计师在新建筑的核心部位做了虚化的处理，使建筑上面的部分架空，这样双方在视轴线上建立概念上的统一，使得新建筑具有历史意识的象征。

② 在新建筑的中央，设计了一个被称之为"协作云"的中空部分，构成建筑的核心，并作为新总部的概念和空间的标识。

③ 空间中设计了标准化且具灵活性的办公空间，以容纳种创造、合作和互动。

Shape Analysis

Formal / Informal

Valley Mirror Window

Formal Office   Informal Office

Layout Analysis

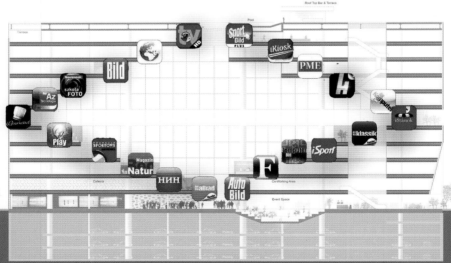

Project Overview / 项目概况

Location analysis
Opposite the existing Axel Springer headquarters on Zimmerstrasse, a street which previously separated East and West Berlin, the building will be situated at one of the city's most significant locations.

区位分析
该项目大楼将坐落于柏林这座城市最具有历史意义的地块之上，恰好在位于西莫大街上的现有 Axel Springer 总部大楼的对面。西莫大街曾经是将东西柏林隔开的分界线。

① The designer adopts virtual-oriented method on the core part of the new building so that the upper part is elevated. In this way, the visual axes of two parts keep in conceptual consistence, giving more historical consciousness to the new building.

② In the new building centre a hollow part named "Coordination Cloud" is designed to act as the building core, as well as a conceptual and space symbol for new HQ.

③ Besides, standard and flexible offices are designed for more creation, cooperation and interaction.

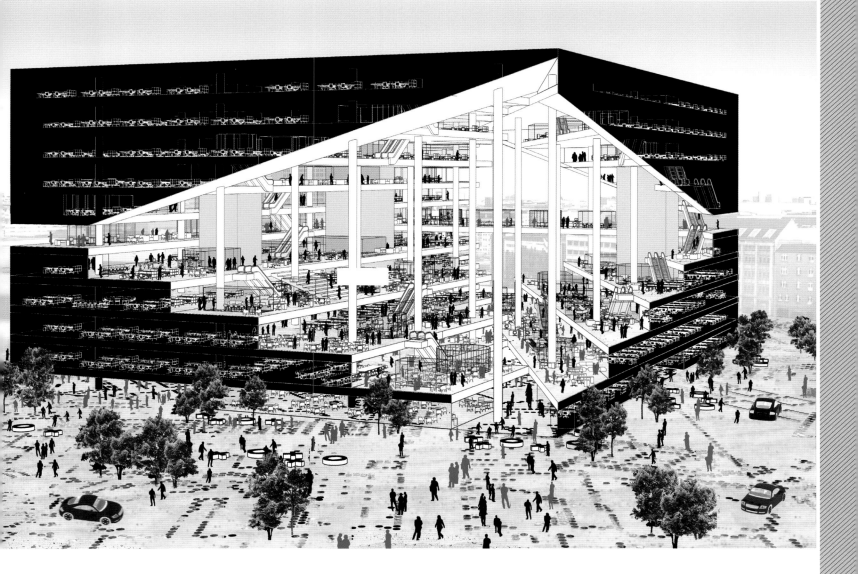

Design concept

The building will embody Axel Springer's corporate shift from print to digital media, and will create a new hub in the existing Axel Springer campus in central Berlin. It presented the conceptually and esthetically most radical model. The fundamental innovation of working environments will support the cultural transformation towards a digital publishing house. On this historical site of all places, for a client who has mobilised architecture to help perform a radical change… a workplace in all its dimensions.

设计理念

该项目大楼的设计将象征着 Axel Springer 公司由印刷业向数字媒体的转变，而且还将在位于柏林中心地区的现有 Axel Springer 总部内创造出一个新中心。Rem Koolhaas 的设计无论从理念还是从美学价值看都是最激进的范例。工作环境的根本创新将推动向数字出版方向的文化转型。该项目客户又非常支持我们在这块颇具历史沧桑感的地块上给建筑大楼设计来一次较为激进的转变，实际上也是对工作空间的各个方面所作的彻底改变。

Plan

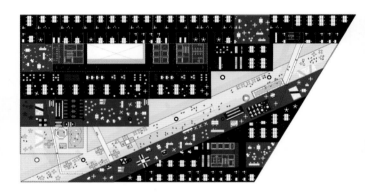

 formal
 informal

Design strategies
The new office block is bisected by a diagonal atrium that opens up to the existing Springer buildings - an extension of the Springer campus. The essence of the proposal is a series of terraced floors that together form a "valley". Each floor contains a covered part for formal work, which is then uncovered on the terraces to act as an informal stage and a place to

设计策略
新办公大楼将被沿对角线分布的中庭分成两大部分。该对角线一直扩展至 Springer 的现有总部大楼——Springer 校区的延伸部分。我们的设计本质是将该项目大楼看作是一系列呈阶梯状分布的楼层叠加在一起共同形成一个"山谷"。每层都

broadcast creative results to other parts of the company. The ground floor is open to the city and contains broadcasting studios, event and exhibition spaces and restaurants.

包含一部分有屋顶覆盖的区域作为传统的工作环境,而剩下的露台未覆盖区域将成为一个非正式的平台以及向公司其他部分传播创新成果的场所。该项目大楼地面第一层向整个城市开放,而且公司的广播室就安排设在这一层,同时还包括各种活动与展览空间以及各种饭店。

PROBLEM / 问题

1. AS A COMPONENT OF THE REDEVELOPMENT OF THE WORLD TRADE CENTRE AND REVITALISATION OF LOWER MANHATTAN

2. RESTRUCTURING MANHATTAN SKYLINE

3. MODERN OFFICE AND MULTI-FUNCTION DEMANDS

4. LOCATED ON BOUNDARIES AMONG AREAS OF DIFFERENT STYLE

1. 为世贸重建和曼哈顿下城振兴计划的一部分
2. 重整曼哈顿天际线
3. 现代办公空间与多重功能需求
4. 位于不同风格的城市区域交界处

2# World Trade Centre
二号世贸中心大楼

BIG
works
BIG
作品

Architect: BIG
Partners in Charge: Bjarke Ingels, Thomas Christoffers
Client: Silverstein Properties & 21st Century Fox / News Corp
Area: 260,000 m²
Location: 200 Greenwich Street, NYC, USA
Function: Commercial office building

设计公司：BIG
合作方负责人：Bjarke Ingels 和 Thomas Christoffersen
客户：Silverstein Properties & 21st Century Fox / News Corp
面积：260 000 平方米
地点：美国纽约市格林威治大街200号
功能：商业办公大楼

IMAGE BY DBOX

DESIGN REQUIREMENTS
设 计 要 求

2 WTC is the capstone in the redevelopment of the World Trade Centre and the final component of the revitalisation of Lower Manhattan. It will serve as the new headquarters for 21st Century Fox and News Corp, which will occupy the lower half of the tower, and more than 5,000 people under one roof.

2 WTC 地块是世贸重建和曼哈顿下城振兴计划的最后一个部分。大楼的下半部分将作为 21 世纪福克斯公司和新闻集团的新总部,届时使用人数将超过 5000。

① 将设计一个高效灵活具有发展潜力的场所,为曼哈顿下城注入媒体新活力。

② 堆叠错位形成的屋顶花园将为人们提供前所未有的露台式城市景观,纽约的天际线因为老式摩天楼与新式摩天楼的交相辉映共同协奏而更加魅力焕发。

③ 大楼通过开放的工作场所,非正式的会议场所以及内部设施提供一个合作与共享的环境。各种空间和谐的联系在一起:篮球场,跑道,餐厅,放映室等等,还有7个天台花园。

④ 七个建筑体量的面积大小不一、深度各异,从面积最大的地面层逐级错落堆叠排列上升至面积最小的顶层,而仿佛由多个大楼组成的整栋大楼,与美国的座右铭"合众为一"有异曲同工之妙。

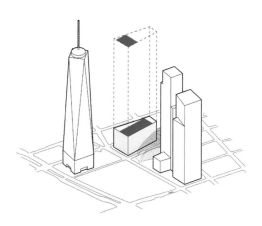

THE SITE
2 WTC is located at 200 Greenwich Street and bounded by Church Street to the East, Vesey Street to the North and Fulton Street to the South. The base of the building utilizes the maximum area of the 56,000 sq ft site.

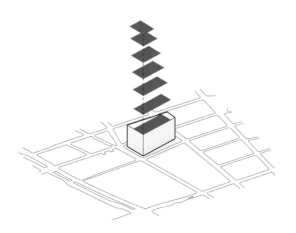

TRANSITION BETWEEN TYPOLOGIES
Floorplates between the maximum-size and minimum-size are optimized to specific tenant needs and requirements.

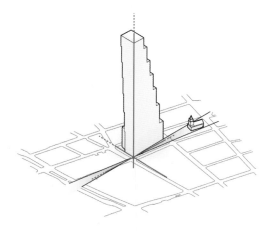

FOLLOWING THE "WEDGE OF LIGHT"
2 WTC is aligned along the axis of Daniel Libeskind's 'Wedge of Light' plaza to preserve the views to St. Paul's Chapel from the Memorial park.

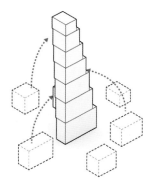

BUILT ON A STRONG FOUNDATION
The needs and requirements of the tenants are concentrated into seven separate building volumes, each tailored to their unique activities. The volumes are stacked on top of each other from the largest to the smallest, creating unity out of diversity.

Project Overview / 项目概况

Location analysis
200 Greenwich Street/2 WTC is the capstone in the redevelopment of the World Trade Centre and the final component of the revitalisation of Lower Manhattan.

区位分析
格林威治大街200号以及2WTC地块是世贸中心重建和曼哈顿下城振兴计划的最后一部分。二号世贸中心大楼座落于格林威治大街200号,东面邻近教堂大街,北面紧靠维西大街,而南面则靠近富尔顿大街。二号大楼建成之后的高度为408.432米,紧邻9•11纪念公园(以示敬意),旁边分别是1号、3号和4号世贸中心大楼。

Design an efficient and flexible place with great potential to render more vitality to Lower Manhattan.

Stepping terraces are heavily planted, creating a vertical succession of the greenery rising. The traditional and innovative skyscrapers add more brilliance to NY skyline.

The new building will provide the physical environment for collaboration and idea sharing through the internal mix of open workplaces, informal meeting spaces, and amenities. ③

The building consists of seven parts of different areas and depths, stacked up to the smallest top layer. Composed of many individual buildings forming a single silhouette, this building appears like an architectural manifestation of the founding motto of this country: "E pluribus unum" - "Out of many, one." ④

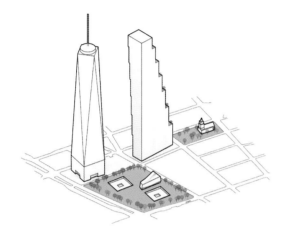

STEPPING TERRACES TO ST. PAUL'S CHAPEL
The terraces are heavily planted, creating a vertical succession of the greenery rising from St. Paul's to the skyline.

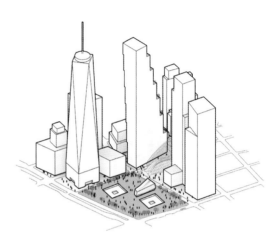

2 WTC
2 WTC is a building that brings together TriBeCa and FiDi at the nexus of the Memorial Park.

 +  =

VERTICAL VILLAGE + MODERN TOWER
2 WTC is located between two different neighborhoods of New York: the residential TriBeCa and the corporate Financial District. The design combines the characteristics of both neighborhoods into an architectural hybrid, melding high-rise with low-rise, modern with historical.

Located at 200 Greenwich Street and bounded by Church Street to the east, Vesey Street to the north and Fulton Street to the south, the tower will rise to 408.432m, respectfully framing the 9/11 Memorial Park alongside 1 WTC, 3 WTC and 4 WTC.

Architectural design

The design of 2 WTC is derived from its urban context at the meeting point between two very different neighbourhoods: the Financial District with its modernist skyscrapers and TriBeCa with its lofts and roof gardens. The design combines the unique qualities of each, melding high-rise with low-rise and modern with historical. From the 9/11 Memorial, the building appears as a tall and slender tower just as its three neighbouring towers, while the view from TriBeCa is of a series of stepped green terraces. The building is aligned along the axis of World Trade Centre Master Planner Daniel Libeskind's 'Wedge of Light' plaza to preserve the views to St. Paul's Chapel from the Memorial park.

建筑设计

2号世贸中心大楼的设计充分考虑了地块所在的城市背景。该地块两边是两片风格非常迥异的区域，它正好处于这两大区域的交汇处——以现代化风格的摩天大楼为特色的金融区以及以阁楼和屋顶花园为特色的特里贝克地区。该项目大楼的设计汲取了这两种风格建筑的独特优点，巧妙地结合了摩天大楼和低层建筑的优点，既不失现代美感，又保留了历史文化的底蕴。从9•11纪念公园的角度看，2号世贸中心大楼仿佛是一座又高又纤细的巨塔（与旁边其他三座世贸中心大楼非常相似）；然而，从特里贝克地区的角度望去，它又仿佛是一系列呈阶梯状分布的绿色梯田。该项目大楼是沿着"光之楔"广场（由世贸中心主设计师丹尼尔•李伯斯金负责设计）形成的轴线排列的，从纪念公园的角度，仍然可以欣赏到圣保罗教堂的风采。

Volume design

The 80-plus storey building will serve as the new headquarters for 21st Century Fox and News Corp which will occupy the lower half of the tower, housing their subsidiary companies and more than 5,000 people under one roof. The upper half of the tower will be leased by Silverstein Properties to other commercial office tenants. The needs and requirements of the media company and other tenants are concentrated into seven separate building volumes, each tailored to their unique activities. The volumes of varying sizes and depths are stacked on top of each other from the largest at the base to the smallest towards the top. The stacking creates 38,000 sf (3,530 sqm) of outdoor terraces full of lush greenery and unprecedented views of the surrounding cityscape, extending life and social interaction outdoors. The modernist skyscraper and the contemporary interpretation of the pre-modern setback merge in a new hybrid and an exciting addition to the NYC skyline.

体量设计

这栋 80 层高的塔楼将成为 21 世纪福克斯公司和新闻集团的新总部办公区所在地——它们将入驻该栋大楼的下半部分楼层,而且其各自的子公司以及超 5 000 位员工都将在这同一栋大楼里办公。同时,兆华斯坦地产公司将向其他商业办公租户出租该大楼的上半部分楼层。媒体公司和其他租户的需求和要求都集中于七个独立的建筑体量,而每一个建筑体量都是根据他们独特的经营活动而特别设计的。七个建筑体量的面积大小不一、深度各异,从面积最大的地面层逐级错落堆叠排列上升至面积最小的顶层。这种错落堆叠的方式创造出一大片户外露台区域,总面积达 3 530 平方米。露台上面不仅种满了郁郁葱葱的植物,而且还是欣赏周边城市风景,拓展生活和社交户外活动的绝佳场所。该项目大楼以一种全新的方式不但展现了现代风格的摩天大楼,而且对前现代时期遭受的挫折作出了当代解读。2 号世贸中心大楼势必将成为纽约城市上空一道令人兴奋的天际线。

Public space design

The base of the building utilises the maximum area of the site, housing TV studios and 100,000 sf (9,290 sqm) of retail space over multiple levels. The 38,000 sf (3,530 sqm) lobby is connected to the WTC transit hub, providing direct access to 11 subway lines and PATH trains. A public plaza at the foot of the building and access to 350,000 sf (32,516 sqm) of shopping and restaurants in the adjacent transportation hub and concourses will ensure life and activity in and around the new World Trade Centre.

公共空间设计

该栋新世贸中心大楼的建成将为人们的协同工作和知识分享提供一个良好的环境,因为其内部设置了各种开放式的,该项目大楼充分利用该地块底层最大面积区域用作电视台公司的办公空间,同时也包括设置多层零售空间(面积达9 290平方米)。3 530平方米的大厅直接与世贸中心交通枢纽站相通,可直接搭乘11条地铁线路和PATH线。该大楼底部建有一个公共广场,人们可直接由此进入面积达32 516平方米的购物和餐饮区,旁边是交通枢纽中心和中央广场。如此一来,不论是在该栋新世贸中心大楼内部还是在其周边地区,生活和工作都非常方便而快捷。

The new building will provide the physical environment for collaboration and idea sharing through the internal mix of open workplaces, amenities and informal meeting spaces. Large stairwells between the floors form cascading double-height communal spaces throughout the headquarters. These continuous spaces enhance connectivity between different departments and amenities, which may include basketball courts, a running track, a cafeteria and screening rooms. The amenity floors are located so they can feed directly out onto the roof top parks.

工作空间、便利设施以及休闲集会空间。在不同楼层之间设置的大型楼梯井形成了呈瀑布状分布的两层高的公共空间，贯通了整个总部办公区。这些连续空间进一步加强了不同部门和设施之间的连通性，包括篮球场、跑道、自助餐厅以及放映室等。同时，该大楼内部还设有各种便利配套设施，在这里工作的人们可直接在屋顶花园享受各种美食。

PROBLEM / 问题

1) HISTORICAL & CULTURAL BACKGROUND
2) LANDMARK
3) MULTI-FUNCTION
4) ENERGY-SAVING

1. 历史文化背景
2. 标志性
3. 多功能性
4. 节能性

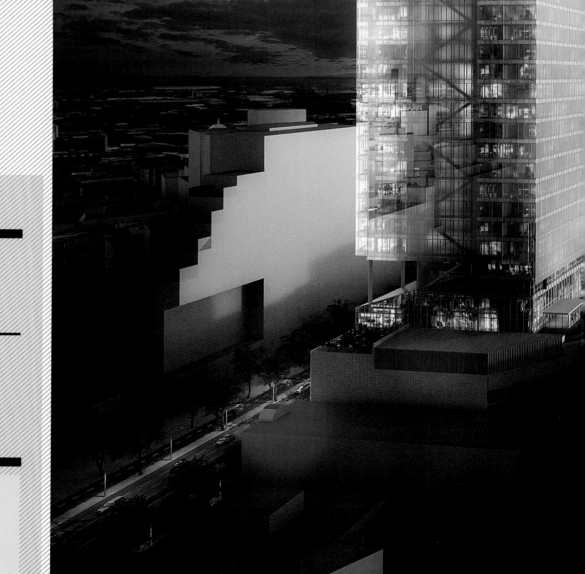

Reforma Towers
改革大厦

Richard Meier & Partners
works

Richard Meier & Partners
作品

Architect: Richard Meier & Partners, Architects LLP
Design Team: Richard Meier, Bernhard Karpf
Local Architect: Diametro Arquitectos
Client: Diametro Arquitectos
Location: Mexico City, Mexico
Area: 120,000 m²
Function: Complex

设计公司：Richard Meier & Partners, Architects LLP
设计团队：理查德·迈耶，伯哈德·卡普夫
当地设计公司：Diametro Arquitectos
客户：Diametro Arquitectos
地点：墨西哥州墨西哥市
面积：120000 平方米
功能：综合体

DESIGN REQUIREMENTS
设 计 要 求

The architects hope this multi-use development project help the reformation of Mexico Centre. It will grow into a new city work and entertainment place.

我们希望这个新混合用途开发的项目有助于墨西哥中心的改革，它将成为一个新的城市工作和休闲活动的场所。

THE SOLUTION
解决方式

① 新改革大厦设计的两座大楼与墨西哥市现有肌理交相呼应、相映成趣；同时也成为了公然反抗传统造型的视觉声明。

② 项目由两座大楼组成，一座是40层的多功能标志性塔楼，设有一系列项目，包括高端办公室、商店、餐馆、一家健身中心及停车场。另一座是27层的酒店大楼。

③ 设计刻意安排一个空间划过体量，从而让建筑结构与服务项目实现再分配，实现别具一格，且高效的配置。

④ 大楼挖空的的姿态保证办公楼内部中心区域自然采光与自然通风的最大化，提高大楼的透明度，同时凸显历史之都和改革大道的美景。

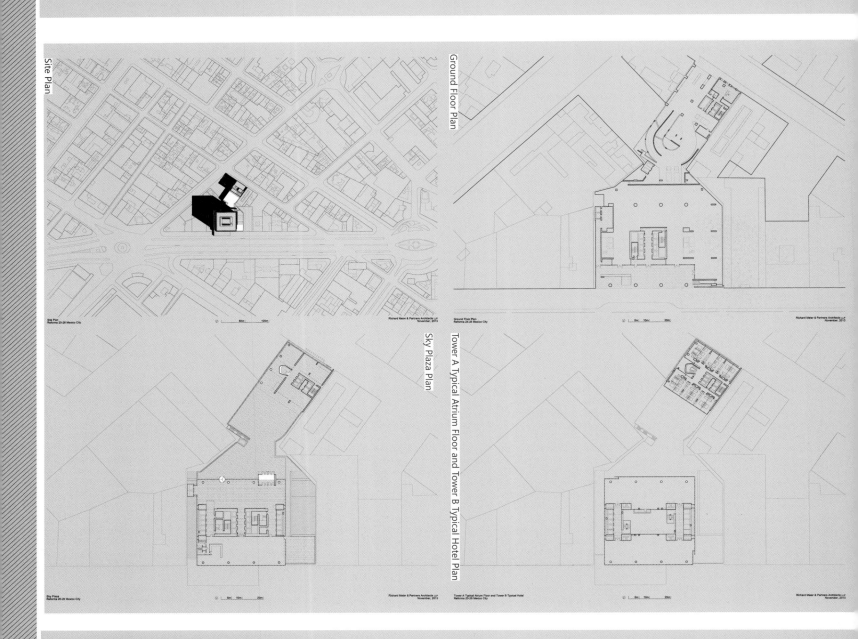

Site Plan / Ground Floor Plan / Sky Plaza Plan / Tower A Typical Atrium Floor and Tower B Typical Hotel Plan

Project Overview / 项目概况

Location analysis
Reforma Towers will be located along Paseo de la Reforma in Mexico City. This distinguished Boulevard was designed to commemorate the history of the Americas and has become a major commercial thoroughfare that cuts diagonally across the city. Sitting boldly along this Boulevard, the proposed development is a mixed-use building complex designed by Richard Meier & Partners.

项目概况
改革大厦坐落于墨西哥市的改革大道。这条著名的林荫大道起初是要纪念美洲的历史，现已成为斜跨整个墨西哥市的一条主要商业干道。坐落在大道旁的多功能建筑综合体非常显眼。综合体设计单位是 Richard Meier & Partners 事务所设计。

1. The two buildings of the new Reform Tower are designed to keep in consistence with the existing Mexico; meanwhile, it also serves as a visual resistance against traditional buildings.

2. The project comprises two buildings. One is a 40-storey multi-function tower, housing high-end offices, shops, restaurants, health club and parking lot; and the other is a 27-storey hotel.

3. The design adopts a space dividing the building so that the structure and service projects are further allocated of different styles, sharing higher efficiency.

4. The elevated part of the building ensures the possibly best sunlight and ventilation in the central area, improves the building's transparency and presenting the harmony between a historical city and the reform road.

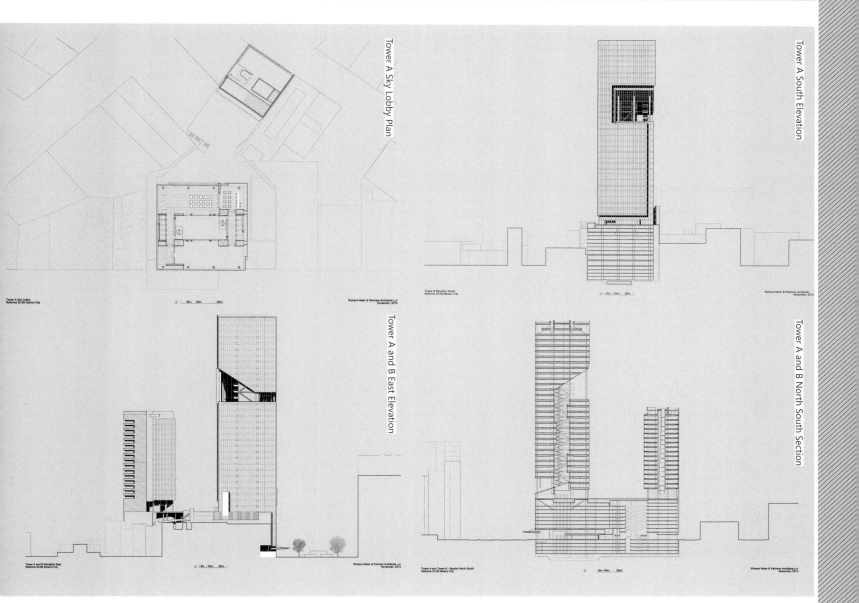

The iconic image of the new Reforma Towers will establish a dynamic relationship between the buildings and the existing fabric of Mexico City; while creating a visual statement which defies traditional tower typologies.

新改革大厦所展现的地标形象将让两座大楼与墨西哥市现有肌理交相呼应、相映成趣；同时也成为了公然反抗传统造型的视觉声明。

Design concept

Mexico City represents one of the most important cultural and commercial centres in Latin America. As the city's economy continues to thrive, it is the architects intention to develop a project that is sensitive to the history of Mexico and its rich architectural legacy.

Design strategies

The new development is comprised of two buildings unified by a base. An iconic 40-storey mixed-use tower that will accommodate a range of programmes; such as high-end offices,

设计理念

墨西哥市是拉丁美洲最重要的文化和商业中心之一。随着墨西哥经济的不断繁荣，我们旨在开发一个彰显墨西哥历史和丰厚建筑遗产的项目。

设计策略

该项目由两座大楼组成，两座大楼矗立在同一地基之上。一座是40层的多功能标志性塔楼，设有一系列项目，包括高端办公室、商店、餐馆、一家健身中心及停车场。另一座是27层的酒店大楼。该大楼采用与第一座大楼同样的设计原理，发挥相辅相成的作用。整个项目的设计综合考虑了城市的局限性、未来的发展及周边环境的变化。

retail space, restaurants, a fitness centre and space for parking. In addition, a 27-floor hotel tower that follows the same design principles as its counterpart will complement the activities of the complex. The overall design of the project considers the current constraints of the city while accounting for the possibility of future development and change of its surroundings. The project's design operations challenge typical tower conventions. By strategically carving a central void through the tower volume, structure and programme become redistributed into unconventional yet efficient configurations. The new possibilities of this internal logic are reflected on the exterior through volumetric cut-outs. A gesture that allows maximising internal natural light and natural ventilation within the centre of the office floor spaces improving transparency and emphasises views of the historic city centre and Reforma Boulevard.

该项目的设计与传统建筑截然不同。刻意安排一个空间划过体量，从而让建筑结构与服务项目实现再分配，实现别具一格，且高效的配置。体量的挖空是表征现象，而隐匿其中的是内在逻辑的创新。大楼的姿态保证办公楼内部中心区域自然采光与自然通风的最大化，提高大楼的透明度，同时凸显历史之都和改革大道的美景。

PROBLEM / 问题

① COMPLEX FUNCTIONS
② EFFECTIVE STREAMLINE
③ INTEGRATION AND SEPARATION OF DIFFERENT FUNCTIONS

1. 功能复杂
2. 流线高效性
3. 不同功能间的融合与分离

RIVM & CBG
Headquarters Building
RIVM & CBG 总部大楼

UNStudio
works
UNStudio
作品

Architect: UNStudio
Client: VolkerWessels Bouw & Vastgoedontwikkeling
Location: Utrecht, The Netherlands
Building Area: ca. 21,000 m²
Function: Research institutes

建筑师：UNStudio
客户：VolkerWessels Bouw & Vastgoedontwikkeling
地点：荷兰乌得勒支
建筑面积：约21000平方米
功能：研究机构

DESIGN REQUIREMENTS

设 计 要 求

RIVM/CBG seek to establish the two organisations on the Uithof in Utrecht. The aim of this integration is to allow for a synergy of universities, educational institutions, research institutes and knowledge intensive companies, as a critical success factor for a climate of knowledge development and knowledge sharing.

RIVM/CBG 的新总部将两个机构设立在了 Uithof 区。这次融合意在让大学、教育机构、研究机构及知识密集型企业拥有协同合作的机会,这是知识发展与分享风气能够成功的决定性因素。

① "十字"设计模型提出将办公区和实验区通过直接的视觉连接,成为一个标准化功能化的整体。

② "中庭主路"则覆盖了十字的中心区域,六层高的恢宏的中心大厅就矗立其中。大厅拥有公共室内广场,它就像一个核心,从视觉上、布局上和流通性上将建筑的四翼组合到了一起。

③ 基底及其以上空间都使用了连续的白色框架进行连接,突出了建筑的功能特点以及外部观感。

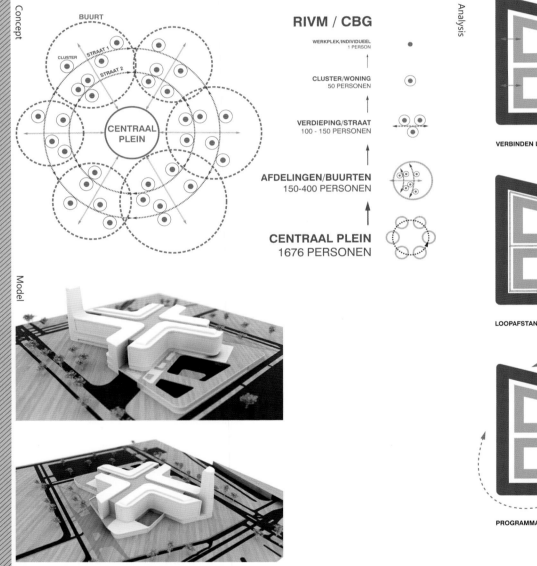

Project Overview / 项目概况

Project overview

The New Headquarters of RIVM/CBG (National Institute for Public Health and Environment and Dutch Medicines Evaluation Board) seek to establish the two organisations on the Uithof, as a part of an open research, knowledge and academic network within the Utrecht Science Park (USP). The aim of this integration is to allow for a synergy of universities, educational institutions, research institutes and knowledgeintensive companies, as a critical success factor for a climate of knowledge development and knowledge sharing.

项目概况

作为乌得勒支科学园(USP)中开放的研究、学问与学院网的一部分,RIVM/CBG(国家公共健康与环境研究所/荷兰药物评价委员会)的新总部将两个机构设立在了Uithof区。这次融合意在让大学、教育机构、研究机构及知识密集型企业拥有协同合作的机会,这是知识发展与分享风气能够成功的决定性因素。

The "Cross" design model proposed would bring together the offices and laboratories into a unified and functional whole, with direct visual connectivity. "Atrium streets" are introduced which converge in the mid.dle of the 'cross', where a 6-storey, spacious central hall is located.

设计理念

This hall is a public, indoor square, a communal core that binds together the four wings of the building visually, organisationally and in terms of circulation.

The base and the space above are connected with continuous white frames, featuring the building functions and external effect.

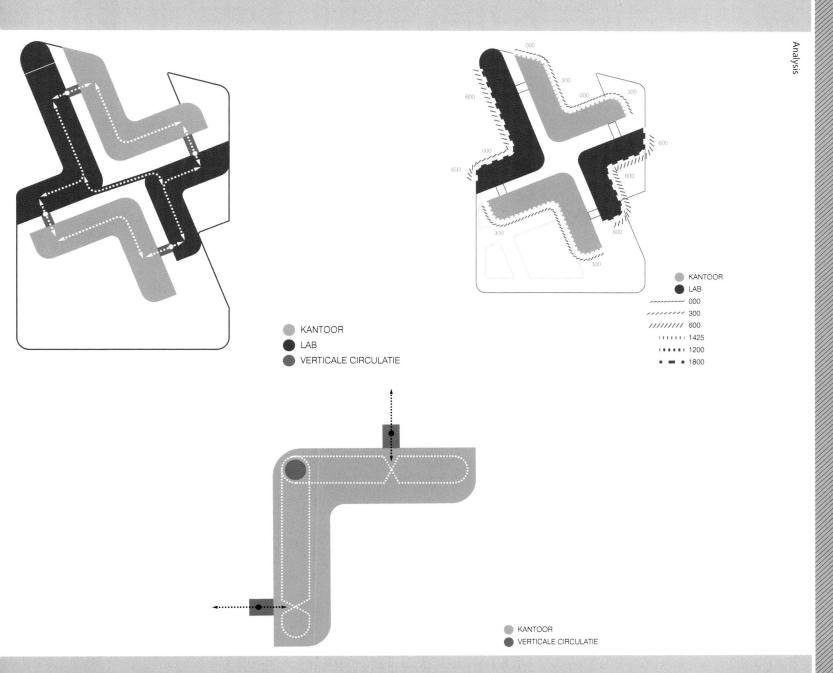

Design concept

The New RIVM uses the "Cross" design model, as well as the facade framing strategy as the "big details" of the building – key design concepts which orchestrate building performance, both to the users inside, as well as the passer-by at Uithof. The Big Detail has the ambition of a highest-rank sustainable strategy, due to its focus on absorbing complex functionalities (spatial, organisational, material, energetic) into a single, multifaceted and effective approach, thus re-defining sustainability as architectural efficiency across all of its conclusive parts.

设计理念

新RIVM大楼使用交叉设计模式，而立面框架设计则作为建筑的"大细节"，成为设计理念中关键的部分，无论对于里面的人还是外面经过Uithof的路人来说，这些立面框架可以说巧妙的规划了整栋建筑。"大细节"志在创造最高级别的可持续战略，它致力于将复杂的多种功能（空间、结构、材料、能源）整合到一个单独、综合而高效的整体中来，通过它的各个组成部分重新定义建筑中能效方面的可持续性。

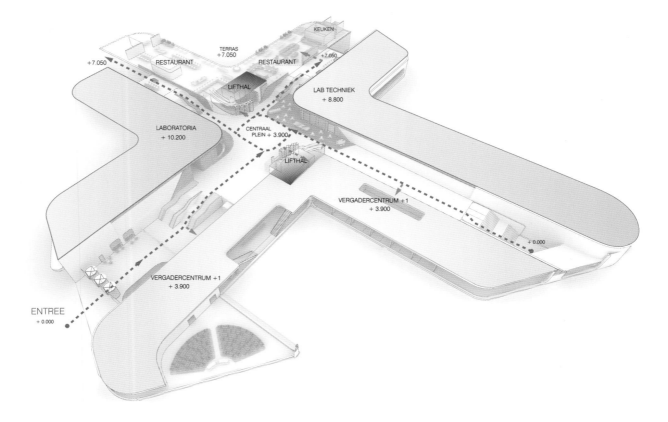

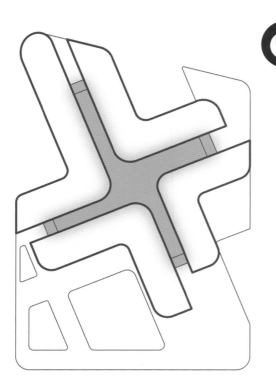

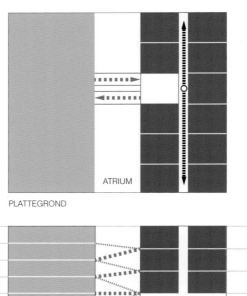

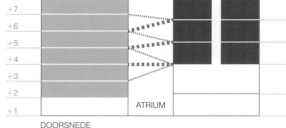

Design strategies
1. Architectural structure
Programmatically, the New RIVM comprises of a two-storey plinth base, housing public amenities (Meeting Centre, Café, Restaurant and Assistance Point) as well as logistics. An undulating volume, combining office and laboratory wings, crowns the plinth. The "Cross" design model brings together these two components of the building into a unified and functional whole, both conceptually and organisationally.

2. Architectura interior
RIVM has a strong base of laboratory functions, with supporting office areas that require direct visual connectivity with the labs. The relationship between offices and labs was therefore designed so that the two functions are brought together via four light-filled voids – the "atrium streets". All four atrium streets converge in the middle of the "Cross", where the 6-storey, spacious Central Hall is located. This hall is a public, indoor square, that binds together the four wings of the building visually, organisationally and in terms of circulation.

设计策略
1. 建筑结构
按照设计，新的 RIVM 两层的基座用于公共设施(包括会面中心、咖啡馆、餐厅和援助点)和物流。起伏的楼体架在基座之上，楼翼用于办公室和实验室。其"交叉"设计将楼体的两部分组合成统一的整体，这样的组合既是概念上的也是结构上的。

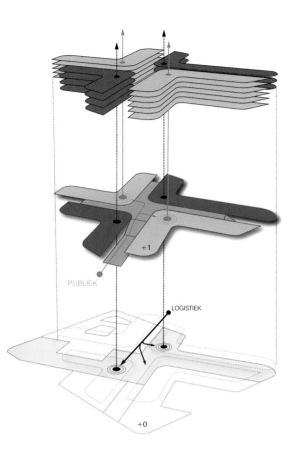
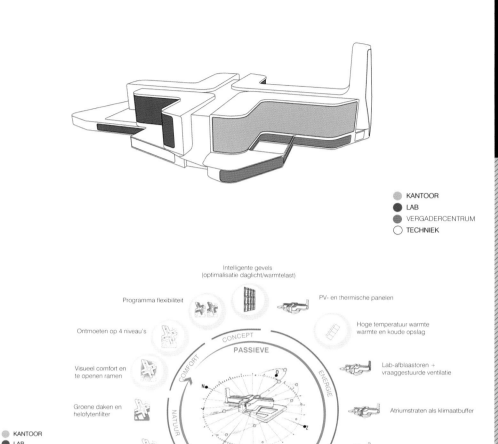
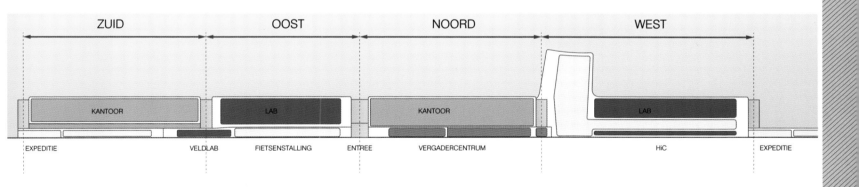

From the Central Hall, which contains the Grand Café and the restaurant, all employees can access their workspaces, whether that be in an office or a lab. Similarly, all logistical circulation of goods and chemicals in the plinth below is brought to a central distribution point and then transported upward. Thus, the centralised organisation of the "Cross" supports both goods and people flow in parallel. The efficiency of this model lies in the shortened travel distance from the central point to the wing ends, which is reflected also in an optimised maintenance solution.

2. 建筑内部

RIVM 为实验室功能的需求提供了良好的基础，实验室的办公区可以直接看见实验室的情况。办公室与实验室之间的联系被设计考虑进去，通过四处明亮的空间——"中庭街"，两者汇聚到一起。这四条中庭街在"交叉"的中心汇合，在那儿坐落着六层高的宽敞的中央大厅，大厅是一个公共的室内广场，通过道路循环流通的方式将四座翼楼粘合在一起，这种粘合是可见的、有组织的。

中央大厅内有咖啡厅与餐厅，无论员工是身在办公室或是实验室，他们都可以从自己的工作进入其中。同样的，下方基座内货物和化学品也会通过物流被带到一个中心集散点，然后再被运送上去。因此，这种"交叉"的中央化结构让货物与人员可以平行流动。从中心去往任何翼楼尽头的距离都被缩短，显示出了这种模式的高效，这种高效也体现在维护方案的优化上。

The Central Hall is the main meeting point of the building, allowing opportunities for different employees to interact informally, to communicate, and to have an overview into the happenings within the building. It is the communaication "core" of the New RIVM, embodying the stewardship of an open workspace, where uninhibited exchange of knowledge, experience and ideas is facilitated. Within the atrium streets, such communication is promoted with the bridges, connecting between the office and the lab wings.

中央大厅是建筑内的主要集合点,为员工创造出私下互动、沟通和就大楼内部各种事件发表看法的机会。这体现了新RIVM的"核心",即开放的工作场所管理,这里便于知识、经验与想法无拘无束的交流。在中庭街里,连接办公室与实验室不同翼楼的天桥也促进了这样的交流。

3. Architectural ventilation and lighting

Connectivity and visual access of the Central Hall is paired with plentitude of daylight - it acts as a guide throughout the building and encourages the healthy use of stairs. Additionally, the centralised "Cross" of voids acts as a climactic regulator of the building, ensuring air exchange and comfort all the way up to the highest floors of offices, open to the atrium.

Formally, the Cross model is reflected in the billowing volume above the plinth, which gently merges with the ground at selected points of the plinth – at the Main entrance, the Logistical entrance and the Lab tower. The tower emphasises the position of the building on the campus, by marking it as a powerful introductory presence on the entrance path toward Uithof. At the same time, this height accent performs as an important sustainable element, offering natural extraction to the laboratory air outlets and saving 80% of the energy on the ventilators.

3. 建筑通风和采光

充足的自然光配合了中央大厅内的连通性与可见的通道,自然光如同向导般指引人们穿梭于建筑内部,并且鼓励他们为健康而使用楼梯。此外,中央化的"交叉"空间还有调节建筑内气候的作用,让建筑从下到上的空间都保持舒适和空气流通。外型上看,这种交叉模式下的楼体如同波浪般起伏地架在底座之上,在底座某些特定的点上与地面逐渐融合,如在主入口、物流入口和实验楼处。实验楼位于去往Uithof的道路入口,那显眼的外观标志了建筑在园区的位置。同时,这座高楼还是建筑可持续性的重要体现,它可通过自然的方式排出实验室的空气,为通风设施节约了80%的能源。

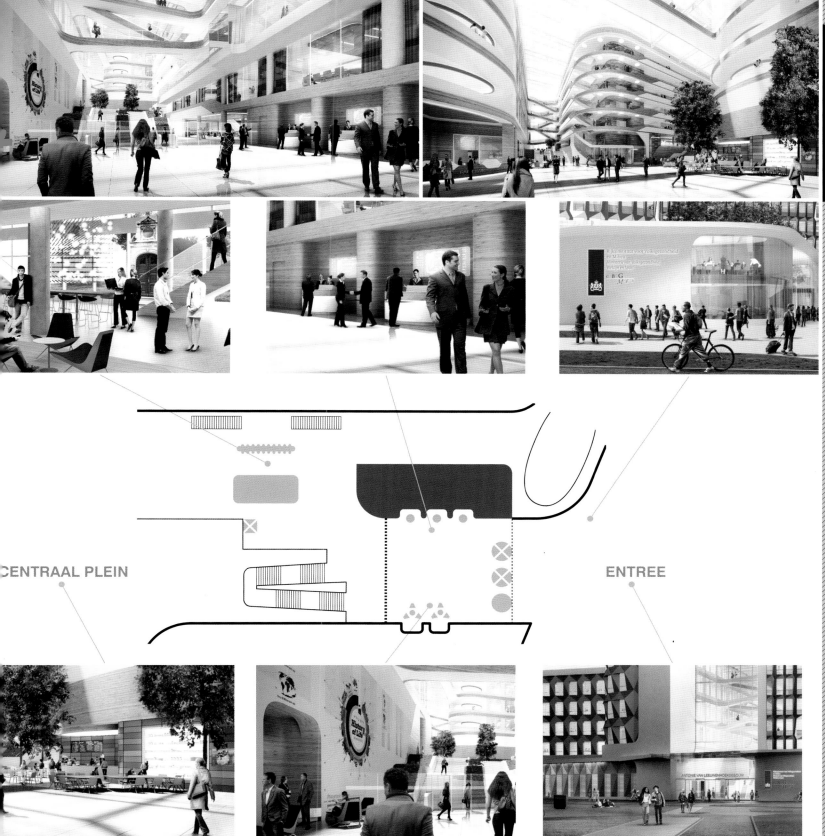

The plinth volume and the volumes above it are bound together by a continuous white framing, which accentuates the programmatic expression of the building, as well as its reading from the outside. The framing is filled with different kinds of "infills" – facade elements which correspond to various programmatic requirements and sun orientation. The laboratory and office facades comprise a number of window components, optimised to respond to heat gain and solar intake conditions. Public programmes within the plinth have more transparent glass facades, whereas the logistical functions are situated behind a fence facade.

底座楼体和上面的楼体由一套整体的白色框架固定在一起，从外部来看，这框架加深了建筑富有规划感的形象。框架内有各种"填充物"，即满足不同规划与朝阳需要的立面。实验室与办公的立面安装了大量的窗户，这些窗户经过优化能更好的满足不同条件下获取热量和日光的需要。底座内的公共区使用了更多的透明玻璃立面，而物流区则用栅栏当做立面。

PROBLEM / 问题

1. **FUNCTION**
2. **ENERGY-SAVING**
3. **SHARE & EXCHANGE**
4. **SECURITY**

1. 功能性
2. 节能性
3. 共享与交流
4. 安全性

Mirai House, Leiden Bio Science Park, The Netherlands

荷兰莱顿生物科学园
Mirai House

UNStudio
works

UNStudio
作品

Architect: UNStudio
Client: G&S Vastgoed
Location: Bio Science Park, Leiden, NL
Site Area: 11,270 m²
Built Area: 3,921 m²
Function: Offices and laboratories

设计公司：UNStudio
客户：G&S Vastgoed
地点：荷兰莱顿生物科学园
占地面积：11 270 平方米
建筑面积：3 921 平方米
功能：办公室和实验室

DESIGN REQUIREMENTS

设 计 要 求

The new laboratory and offices for the Japanese firm Astellas houses both offices and laboratories. Essential to the design of the building is the creation of a pleasant, open and transparent working environment for Astellas employees, in addition to an agreeable and welcoming gesture to their international visitors.

日本公司 Astellas 会将其办公室和实验室设在新建的实验室和办公区内。建筑设计旨在为 Astellas 员工创造一个舒适、公开和透明的工作环境，同时也能更愉悦、热情地接待国际访客。

① 建筑框架集成了三个不同高度的建筑体，围合成一个内部庭院花园。西边的六层大楼是办公空间。区域的稍低楼层则设有一个带屋顶露台的餐厅，东边的四层大楼则用作实验室。

② 设计通过立面玻璃系统减少内部的热负荷及地下能源存储系统减少能源的使用。

③ 建筑采用开放式的结构搭建，透明的环境空间有助于员工之间的沟通交流。

④ 设计时充分考虑到了安全这个基本因素，通过开放的外墙结构体系集成安全的概念。

Concept

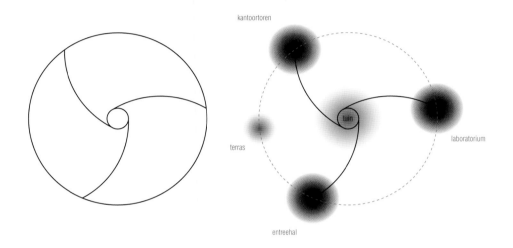

Analysis 1

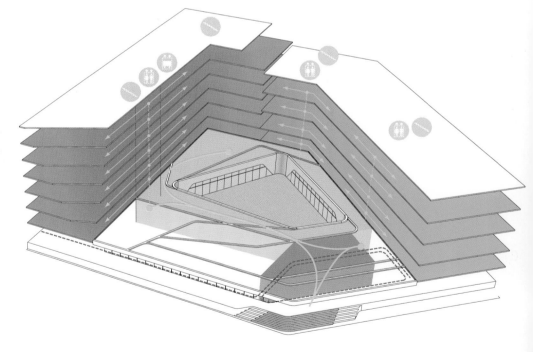

Project Overview / 项目概况

Design strategies

Security is an essential element in the design of the building, which houses both offices and an ultramodern laboratory for scientific research. Rather than approaching this issue in an exclusive way - by confining and fortifying the structure - the design for the Astellas building integrates the security concept inclusively, with the frame of the building serving as an unconcealed enclosure.

The building frame consists of three buildings in different heights, to form an internal garden. The six-storey building on the west is used as offices. The lower building is equipped with a restaurant roofed and the four-storey building on the east is used as labs.

The design successfully reduces internal heat load through vertical glass system and saves energy through energy storage system.

The building adopts an open structure, with transparent space helpful for communication among employees.

Security is fully taken into consideration during design. Security is thus ensured through an open external wall structure.

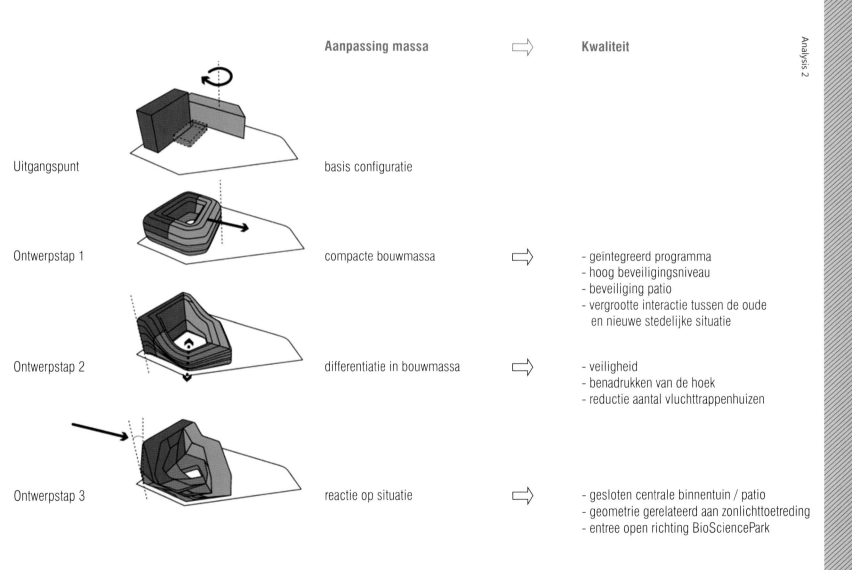

设计策略
这栋建筑包括办公室和超现代的科学研究实验室，设计时充分考虑到了安全这个基本因素。stellas 的设计没有通过封闭和强化结构的方式来解决这个问题，而是在开放的外墙结构中集成安全的概念。

View Analysis

Sunlight Analysis

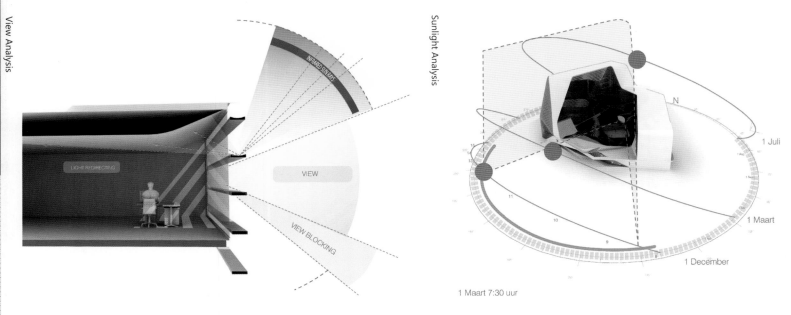

1 Maart 7:30 uur

Section BB

doorsnede B-B

Section Details

1 Aluminium Cladding
2 Sedum Roof
3 Natural Stone
4 Concrete Stair
5 Glas Facade
6 Stainless steel Balustrade
7 HPL ceiling

Equally essential to the design of the building is the creation of a pleasant, open and transparent working environment for Astellas employees, in addition to an agreeable and welcoming gesture to their international visitors. The organisation and materialisation of the building ensures clear views from each of the three areas within the main frame. Glass facades are employed to provide sufficient daylight, whilst also creating open visual communication throughout the structure.

The building frame integrates into one gesture three sections of varying heights, which together encircle an inner courtyard garden. Covering six floors on the west side of the building are the office spaces. A restaurant with a roof terrace is located on the lower floor of this area. The eastern section of the building houses four floors of laboratories.

The main entrance to the building is located on the Southast East and employs the concept of the hotel lobby to comfortably welcome employees and visitors from home and abroad. The floor-to-ceiling glass facades and large skylights of the expansive, 3.7-meter-high lobby area provide visual links to both the inner garden and the surrounding street life. The western side of the lobby area houses seating areas and meeting rooms. Direct access to the inner garden, which is based on traditional enclosed Japanese gardens, is possible from all three sections of the building.

在建筑设计工程中，同样必不可少的考虑因素是为 Astellas 员工打造愉快、开放和透明的工作环境，并为其国际访客打造令人愉快的迎宾环境。建筑的组织和物质结构确保可以从主框架的另外三个区域获得清晰的视图。建筑采用玻璃外墙来提供足够的日光，同时在整个结构中打造开放的视觉联系。
建筑框架集成了三个不同高度的建筑体形成，围合成一个内部庭院花园。西边的六层大楼是办公空间。而此区域的稍低楼层则设有一个带屋顶露台的餐厅。东边的四层大楼则用作实验室。
建筑的主入口位于东南方，采用了酒店大堂的设计概念，旨在为员工和国内外访客打造舒适的迎宾空间。高大天花板的玻璃外墙和 3.7 米高的大堂中的巨大天窗为室内花园和周围的街景提供了视觉联系。大堂西部设有座位区和会议室。内部花园采用传统封闭日式花园，从建筑的三个区域都可以直达这个花园。

Parking is provided at street level to the east of the building and in a sunken parking garage underneath the main structure. The ground floor of the building is raised to a height of 1.7 metres and is accessed by steps which lead to the lobby area from street level. The floor plans in the interior are flexible and based on the campus concept, where emphasis is placed on communication.

建筑西部的沿街区域设有停车场，主建筑的地下层还设有下沉式停车场。建筑的底层抬高至 1.7 米，可经台阶从街边进入大堂区域。内部空间采用灵活的地面规划和校园式设计理念，并重点强调沟通的概念。
Astellas 的立面采用玻璃和铝材作为主要材料，旨在用最少的材料打造轻量级结构。每层立面都采用薄板来转移阳光直射。另外，覆盖在外层的玻璃窗格进一步减少了内部的热负荷。建筑北部的阳光照射较少，所以采用高大天花板的玻璃墙以及 4 厘米厚的薄板。南部和西部 / 东部的立面分别装饰有 30 厘米和 20 厘米厚的薄板。这些立面还采用 90 厘米厚的护栏，进一步的减少阳光直射。整栋建筑的总容积率为 89%-90%。地下能源存储系统可减少能源的使用。日本动漫电影的四个截然不同的色调形成的垂直变化让立面的颜色变幻多端。

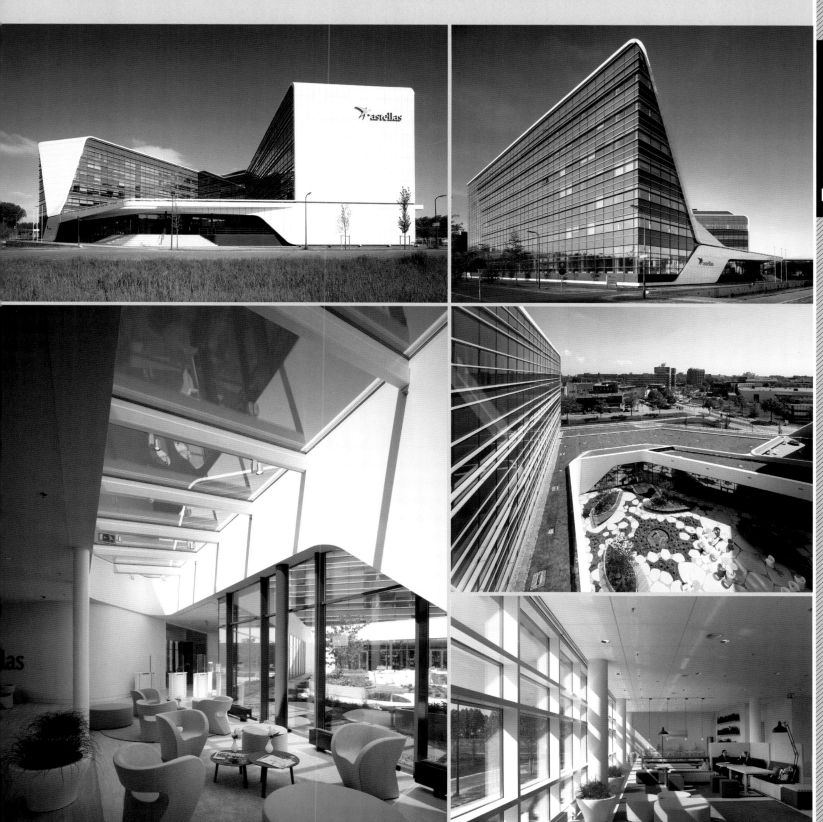

Glass and aluminium are the main materials used in the facade of the Astellas building in order to provide a lightweight structure which requires less material usage in its foundations. Lamellas are incorporated on each level of the facade in order to deflect direct sunlight. The glass panes are further coated to additionally reduce heat load to the interior. On the Northern side of the building, where sunlight is less prevalent, floor-to-ceiling glass is employed, with 4cm deep lamellas. The south and east/wrt54est facades respectively are furnished with lamellas with a depth of 30cm and 20cm. Parapets of 90cm are also employed on these facades in order to further reduce direct sunlight penetration. The building as a whole has an 89%-90% net to gross floor ratio, with an underground energy storage system further reducing energy usage. Colour is introduced into the facade by means of a vertical variation in the four contrasting tones of the Japanese Manga animation films.

PROBLEM / 问题

1. FUNCTIONAL AREAS AND WORK ENVIRONMENT IN DISORDER

2. TWO INDEPENDENT OFFICE PLACES

3. REINFORCED STRUCTURE

1. 功能区域和工作环境凌乱
2. 两个独立的办公场所
3. 结构加固

Blizzard Headquarters
BLIZZARD 办公总部

REX works
REX 建筑事务所
作品

Architect: REX
Client: ACTIVISION | BLIZZARD
Location: Santa Monica, California, USA
Area: 13,300 m²
Composition Space: Including open and private offices, conference rooms, gaming areas, a screening room, an all-company assembly space, and a cafeteria
Function: Office building

设计公司：REX 建筑事务所
地点：美国加利福尼亚州圣塔莫妮卡
面积：13,300 平方米
空间组成：互动娱乐软件行业领先的出版商的办公总部大楼，包括公共以及私人办公室、会议室、游戏区、放映室、全公司集会空间以及自助餐厅
功能：办公大楼

DESIGN REQUIREMENTS
设计要求

Transformed a 1970s office building into a more efficient work space with new amenities for the employees of interactive entertainment company Activision/Blizzard, including open and private offices, conference rooms, gaming areas, a screening room, an all-company assembly space, and a cafeteria.

将一座二十世纪七十年代的办公楼改造成一个更高效的工作场所，为互动型娱乐公司Activision/Blizzard员工提供各种新型便利设施。该建筑物涵盖一个公开和私人办公区、会议区、游戏区、放映室、全公司内部集会区以及一个咖啡馆。

① 设计消除了之前多租客使用设计而形成的间隔，将原来的小隔间改成更高效地工作室储存和办公空间不变，使用面积却减少了 25%。

② 小面积干预取代建筑物原图 8 方案中的双倍宽度的中央波纹板。

③ 中间的区域结构被改造成支撑整座大楼的框架。

Concept

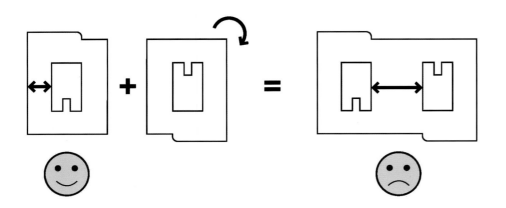

Design of the original building's plan began as a donut with a good width for offices. However, the donut was then duplicated and rotated, resulting in a figure-8 layout with an ill-proportioned middle.

该栋大楼的建筑规划最初设定为将整栋大楼建设成一个"圆环"形，每个办公室的空间都较为宽阔。但是，一个"圆环"的设计后来变成了两个；最终整栋建筑的外观布局呈现出数字"8"的形状，中间部分则极其不匀称。

Floor Plan

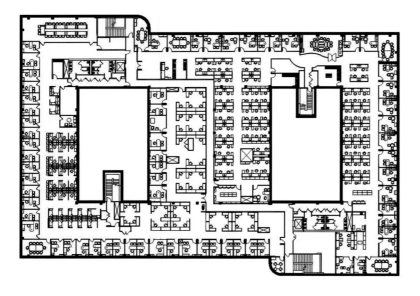

Moreover, the building was originally multi-tenant. A Activision|Blizzard grew, it annexed portions of the structure unt it occupied the building's entirety. Vestiges of the multi-tenan subdivisions remain, creating a confused working environment.

此外，该栋大楼最初是由多个租户共同租赁办公的。随着 Activision | Blizzard 公司的发展壮大，它逐步将该栋大楼的其他租户所持有部分合并，并最终独自拥有了这整栋大楼。然而，多租户分租划分区域的痕迹尚在，整栋大楼的工作环境并不清晰明了。

Project Overview / 项目概况

Design goals

Within this existing structure, Activision|Blizzard aims to:
1. Increase company collaboration, community, and happy accidents.

设计目标

在现有大楼结构的基础上，Activision|Blizzard 公司期望达成如下目标：
1. 增加公司协作凝聚力、团体合作力以及创造与快乐工作生活有关的故事；

The design removes partitions remaining from previous multi-tenant use, and exchanges cubicles for efficient work stations with the same amount of storage and desk space, while occupying 25% less area. ❶

The faceted intervention replaces the double-width central floor plates in the original building's figure-8 plan. ❷

The central structure is reformed into a frame supporting the whole building. ❸

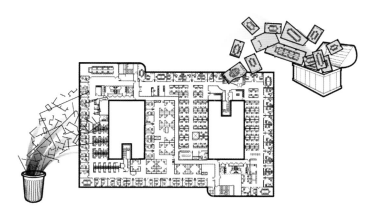

To improve efficiency and rigour of the perimeter Machine, all vestiges of multi-tenant occupation are eliminated and all collective spaces are aggregated, to subsequently form the collaborative Nucleus.

为了提高利用效率以及改善外缘 Machine 办公区的刻板布局，所有的多租户划分区域残留痕迹皆已清除，所有的集体空间皆已合并，并最终形成了具有相互协作功能的 Nucleus 区。

The entire current staff of 519 — plus growth for 114 future employees — fit efficiently on Levels 1, 2, and 3 of the perimeter Machine alone. Large and extra-large offices are grouped in the plan's corners to increase leadership synergy. Inefficient cubicles are replaced with efficient workstations that provide the same amount of work and storage space in 25% less area. Increased efficiency allows for the addition of collaborative desks and storage throughout the open workstations.

Activision|Blizzard 公司目前总共拥有 519 位员工，未来可能需再增加 114 位员工。仅在外缘 Machine 部分的第一、二和三层就可以有效地将这些员工安置妥当。根据相关设计规划，大型和超大型办公室将集中安排在各个角落区域，以增强领导协同作用。低效率小隔间将被高效工作站取代，即使在面积减少 25% 的情况，仍然能够完成相同的工作量，留置相同的存储空间。空间使用效率的提高为在整个开放型工作站内增加协作部门和存储空间创造了有利条件。

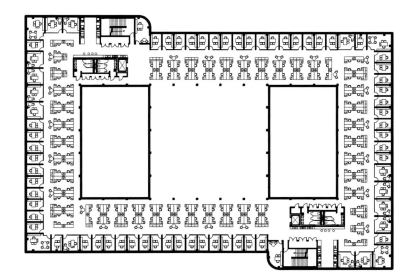

2.Give back to its staff by resurrecting existing amenities – such as courtyards – and providing new amenities – such as a cafeteria, gaming areas, a screening room, and an all-company assembly space.
3.Celebrate the company's staff, the work they create, and their audience.
4.Support Activision|Blizzard and Activision Publishing as separate entities and as a cohesive whole.

2. 为了员工更好的工作生活，重新改造现有便利设施，比如庭院休闲区，并提供新的便利设施，包括自助餐厅、游戏区、放映室以及全公司机会空间；
3. 展示公司员工的风采，表彰他们所做贡献以及他们俘获的支持者；
4. 突出 Activision|Blizzard 公司和 Activision 出版社既是各自独立的个体，也是相互联系的整体。

Due to the Machine's efficiency, the existing collective programmes can be enlarged and new amenities - such as a cafeteria, gaming areas, a screening room, and an all-company assembly space - can be added to the overall building. These collective programmes are amalgamated into the figure-8's middle to form the Nucleus.

由于 Machine 办公区空间的高效运用，现有的集体项目可以适当扩展并在整栋大楼之中增加新的便利设施，比如自助餐厅、游戏区、放映室以及全公司集会空间。这些集体项目都已集中并入数字"8"的中间位置，已形成 Nucleus 区。

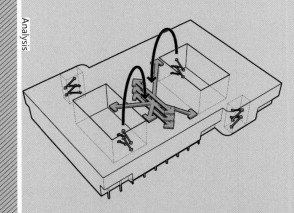

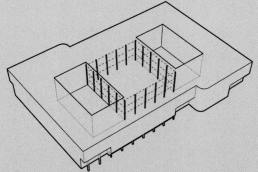

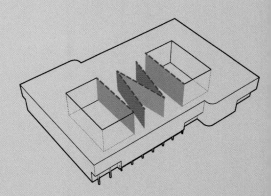

Vertical circulation in the figure-8 is currently dispersed and movement across the plan is difficult. By reconstituting the redundant courtyard egress stairs located in the Nucleus, company cohesion and serendipitous interactions are further fostered.

目前，数字"8"垂直方向的人流已被分散开来，跨越平面布局的活动也是非常困难的。通过重新改造位于 Nucleus 区多余的庭院出口楼梯，公司的凝聚力和偶发性互动将得到进一步的加强。

The building's structural constraints provide an opportunity for determining the Nucleus' architectural solution. Following the 1994 Northridge earthquake, the structure circumscribing the figure-8's middle was retrofitted to create a moment frame that laterally supports the building. The floor plates of this central area are an integral part of the moment frame; any cuts into them would compromise the entire structural system. Tearing out Levels 1, 2, and 3 within the moment frame and rebuilding them is therefore cheaper than adaptively reusing the existing floors.

整栋大楼的结构性限制为最终决定 Nucleus 区的建筑解决方案提供了机会。自1994年的北岭大地震以来，限定数字"8"中间范围的组织结构得到了重新改造，形成了一个起关键性作用的框架，可在侧面支撑整栋大楼。中心区域的楼板构成这个关键性框架不可或缺的一部分；任何切割楼板的行为都将削弱整个组织结构体系的支撑作用。因此，与重新利用现有楼层相比，拆除这个关键性框架的第一、二和三层并重建它们所耗费的成本更低。

Once the existing floor plates of the figure-8's middle are removed, the new Nucleus must reinforce the moment frame by transferring loads in the North-South direction and supporting the columns from buckling in the East-West direction. The most direct way to do this is by providing stringers that connect the moment frame's columns in an "N" configuration as well as along the sides of the courtyards.

数字"8"中间区域的现有楼板一旦被拆除，新增的 Nucleus 区必须加固其起关键性作用的支撑框架，转移南北方向的负重并且支撑起沿东西方向放置的立柱。达到此目的最直接的方法就是使用纵梁按照字母"N"的形状并沿着庭院的侧边，将关键性框架的立柱逐一连接起来。

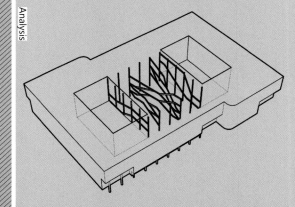

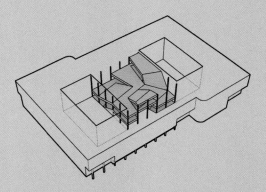

While banal in plan, the stringers can be bent in section to efficiently transfer North-South lateral loads to the ground and provide freedom to solve the geometries of the circulation and programme.

按照常规的平面布局方法，可将纵梁部分弯曲，有效地将南北侧边方向所承受的负重转移至地面，并为解决循环流动部分的几何结构布局提供自由空间。

When topped with floor slabs, the stringers become "folded plates" that accommodate the Nucleus' programme and provide the necessary paths for direct circulation.

当拆除楼板时，纵梁将起到"折叠板"的作用，以适应 Nucleus 区的规划设计，并且为直接循环流动提供必要的路线。

The Nucleus Design Steps

Combining the structural requirements with the programme and circulation requirements makes a very clear set of rules for how the Nucleus should be determined.

Nucleus 区设计步骤
通过将相关的结构要求和项目设计以及循环流动规划要求综合起来考虑,我们制订了一套关于应何如何最终确定 Nucleus 区的非常清晰的规则。

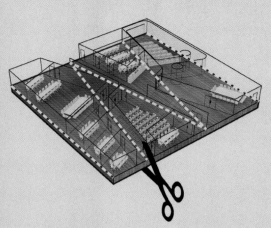

Step 1: Place all enclosed collective programme within a box on Level 2 of the figure-8's middle, and make cuts along the stringer lines.

步骤1:将所有的封闭型集体项目安排设置在数字"8"中间部分第二层的一个箱形空间中,并沿着纵梁线进行切割。

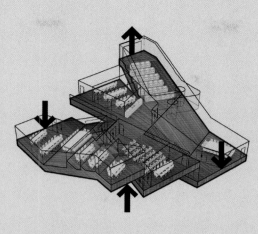

Step 2: Fold two side arms down to make circulatory connections to Level 1, and fold two side arms up to make circulatory connections to Level 3.

步骤2:将两条侧臂向下折叠,以便与第一层进行循环连接;将两条侧臂向上折叠,以便与第三层进行循环连接。

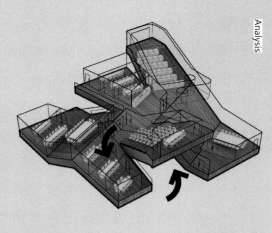

Step 3: Switch one side arm with a centre arm to make a more grand entry condition on Level 1.

步骤3:将一条侧臂与一条中臂进行交换,使得第一层的入口条件更加完善。

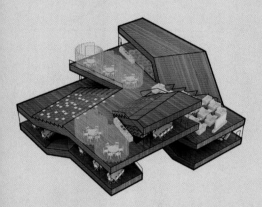

Step 4: Solidify the box's roof to become a public "terrain" of unenclosed collective programmes, and to complete the desired connections.

步骤4:固化箱形空间的屋顶,使其变成非封闭型集体项目的一个公共"地带",并完成项目所需的连接。

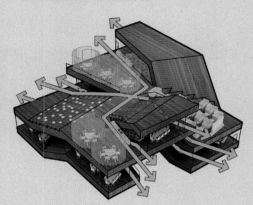

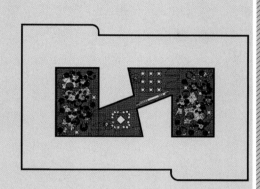

Both courtyards, the cafeteria, and a gaming space are combined to create a single, large amenity for the entirety of Activision|Blizzard on the ground floor.

为了展现位于第一层的 Activision|Blizzard 公司的整体感,庭院、自助餐厅以及游戏区都已被组合起来,作为一项单一大型的便利设施。

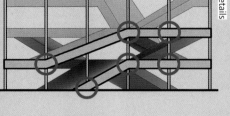

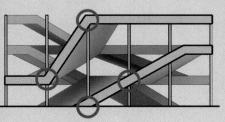

By aligning stringer bends with column intersections, the East-West columns of the moment frame are supported from buckling.

纵梁弯曲部分应与立柱的横截面排成一行,关键性框架东西方向的立柱应从纵梁的弯曲部分获得力量支撑。

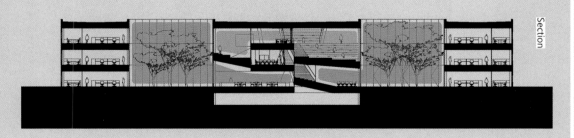

Public Realm
In addition to methodically solving the programmatic, circulatory, and structural requirements, this strategy defines a public realm on Levels 2 and 3 that improves connectivity within Activision Publishing, and a public realm on Level 1 for the entirety of Activision|Blizzard.

公共区域
除了能够有条不紊地满足了该项目计划、流动以及结构限制的要求之外,该项策略性的解决方案还在第二和第三层之间划定了一个公共区域,以完善 Activision 出版社室内的连接性;同时也在第一层划定了一个公共区域,以提升 Activision|Blizzard 公司的整体感。

PROBLEM / 问题

① <u>ENTRANCE RESTRUCTURING</u>
② <u>ICONIC BUILDING</u>
③ <u>CULTURAL HERITAGE</u>

1. 入口空间改造
2. 标志性建筑
3. 文化遗产

Battersea Power Station Malaysia SQUARE

巴特西电站马来西亚广场

BIG
works

BIG
作品

Architect: BIG
Partners in Charge: Bjarke Ingels, Andreas Klok Pedersen, Kai-Uwe Bergmann
Client: Battersea Power Station Development Company
Location: London, UK
Area: 3,500 m²
Function: Public space

建筑公司：BIG
合作方负责人：Bjarke Ingels, Andreas Klok Pedersen, Kai-Uwe Bergmann
客户：Battersea Power Station Development Company
地点：英国伦敦
面积：3 500 平方米
功能：公共空间

DESIGN REQUIREMENTS
设计要求

During the contest, architects, landscape designers and urban planning designers are expected to design a space for the citizens to memorise the relation between Malaysia and UK in the history so that the name of Malaysia will be permanently remembered in London centre. Besides, they should regard this project as the revitalisation of a symbolic station.

设计竞赛希望建筑师，景观设计师和城市规划师设计一个公民的空间，纪念马来西亚和英国之间的历史关系，使得马来西亚的名称被永久设置在伦敦的中心，并作为识别标志性电站建设的复兴工作。

THE SOLUTION / 解决方式

① 广场的设计，灵感来源于马来西亚的姆鲁国家公园内发现的洞穴，极具马来西亚景观和地质特色。

② 两层的公共空间在创造出一条沉静而优雅的入口大道的同时，帮助引导游客们直达巴特西公园站。

③ 将景观、建筑、城市以及媒体的设计相互融合在一起，使得新的马来西亚广场作为雄伟的工业遗产赢得更多的尊重。

INTERACTIVE WATER FEATURE
A playful water fountain is installed on the main agora plaza, controlled by sensors that capture the movement of passing pedestrians. Water disappears when the visitors get close so that it is possible to walk through the fountain with dry shoes.

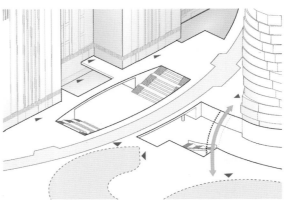

CONNECTIONS
The layout of the previous proposal offers reasonable flows between the main functions of the plaza, except between the plinth of the two sides of the Electric Boulevard. Therefore we have decided to respect the previous scheme and only add a new pedestrian bridge to connect the Electric Boulevard with the Prospect corridor on the level +8.00m to optimize the flows.

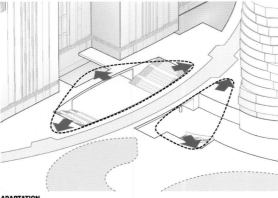

ADAPTATION
The giant landscape openings are larger and more spacious, maximizing the views and presence of the Power Station. Thus the monumentality of its southern elevation, wash towers and chimneys can be even better appreciated.

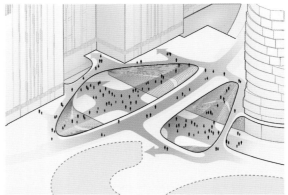

FORM FOLLOWS FLOW
The architectural concept is born out of a profound urban analysis of this magnificent site in front of the iconic Battersea Power Plant, as well as an analysis of the functional requirements. The wavy forms of the steps follow the most efficient layout of the flows.

Project Overview / 项目概况

Upper level planning

Silicon Valley has been an engine of innovation driving technological evolution and global economy. So far the majority of these vast intellectual and economical resources have been confined to the digital realm -- Google North Bayshore expands this innovative spirit into the physical realm.

The square design is inspired by caves discovered in Gunung Mulu National Park of Malaysia, with typical Malaysian landscape and geological features. ①

The two-storey public space creates a quiet and elegant entrance avenue, and helps tourists to Battersea Park Station. ②

The design of landscape, building city and media is perfectly integrated with each other, earning more respect to new Malaysian Square as a great industrial heritage. ③

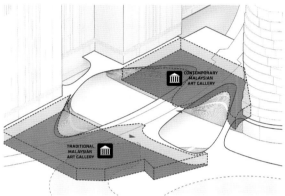

INTERIOR PROGRAM
The main pedestrian route on level +3.00m towards the Battersea entrance is flanked on both sides by galleries dedicated to Malaysian art.

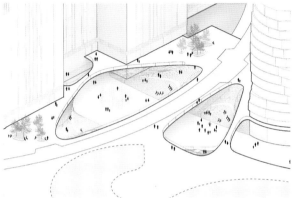

PUBLIC SPACE
In front of the main entrance of the Power Station, a main event space, agora, is carved into the plinth, providing a platform for activities, as well as the ever-changing water installation.

DEMOLISHED CHIMNEYS
The demolished chimneys of the power station would be reclaimed and become part of the aggregates of the floor materials on the plaza. The two countries - UK and Malaysia - are thus bonded together in the Malaysia Square.

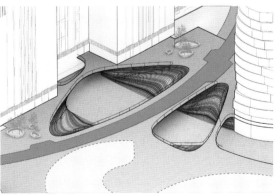

MALAYSIAN GEOLOGY
When walking up the stairs, visitors can experience all the different geological layers of Malaysia.

上层规划
美国硅谷已然成为创新的引擎，不断推动着技术革新和全球经济的发展。目前，大部分丰富的知识和经济资源仍然仅限于数字领域——而谷歌 North Bayshore 将把这种创新精神扩展至物理领域。

 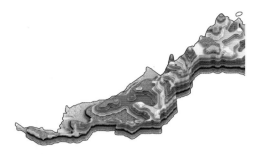

MALAYSIAN REFERENCE - GEOLOGY
The square will be finished in limestone, granite, marble, sandstone, gravel and dolomite, all stone types that are found within Malaysia's geology. The stones are organized in accordance to each quarry's original altitude, becoming a concentrated sample of Malaysian geology.

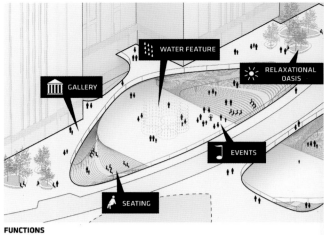

FUNCTIONS
Malaysia Square is designed not merely as a space to move through but also as a space to spend time in – it is a landscape of possibilities.

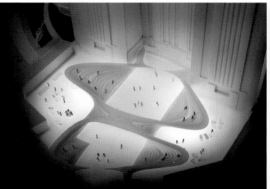  

Design concept
BIG together with Heatherwick Studio and Google have set out to imagine the work environments of future Googlers to be as adaptable, flexible and intelligent as the rest of Google's wide spanning portfolio – rather than an insular corporate headquarters, Google North Bayshore will be a vibrant new neighbourhood of Mountain View.

设计理念
BIG 将与英国托马斯·赫斯维克工作室以及谷歌通力合作设计谷歌新总部。他们已经开始构想未来谷歌人应该身处何种工作环境。它应该像谷歌其他的大跨度设计作品一样,具备适应性强、灵活度高以及智能型设计的特点——而不仅仅只是一个封闭的公司总部。谷歌 North Bayshore 新总部园区将成为美国山景城一道全新的、充满活力的风景。

PROBLEM / 问题

① **CITY BACKGROUND**
② **LANDMARK**
③ **FUNCTION**
④ **ENERGY-SAVING**

1. 城市背景
2. 城市地标
3. 功能性
4. 建筑节能

Bankmed Headquarters II
BankMed 总部 II

Morphosis works
Morphosis 作品

Architect: Morphosis
Client: BankMed/Med Properties
Location: Beirut, Lebanon
Area: 16,000 m²
Function: Office building

设计公司：Morphosis
客户：BankMed/Med Properties
地点：黎巴嫩，贝鲁特
面积：16 000 平方米
功能：办公大楼

DESIGN REQUIREMENTS
设 计 要 求

Beirut is under the developing reconstruction. It calls for a city landmark to unify Beirut to create the wealth of a new Beirut.

贝鲁特地区正处于发展中的重建时期,急需一个城市地标让贝鲁特团结起来,来重新发掘和重写贝鲁特城市的财富。

THE SOLUTION / 解决方式

① 新的 BankMed HQII 被构想为两支彼此分离又相互联系的竖着的长笛,伸向天空,预示着大楼将成为现存街区和未来开发区之间的城市枢纽。

② 现场天然的有利地点、枴角条件、轴线和定位以及独特的建筑造型为建筑大楼的地标形象打下了基础。

③ "双笛"构想让大楼一分为二,一半用于容纳各种功能性的设施和服务,另一半容纳办公室、公共区域和零售银行。

④ 双塔楼的内部有一套外壳系统,包括东塔单层玻璃上带反射的双立面,和西塔的玻璃拱肩立面,以达到建筑整体节能的目的。

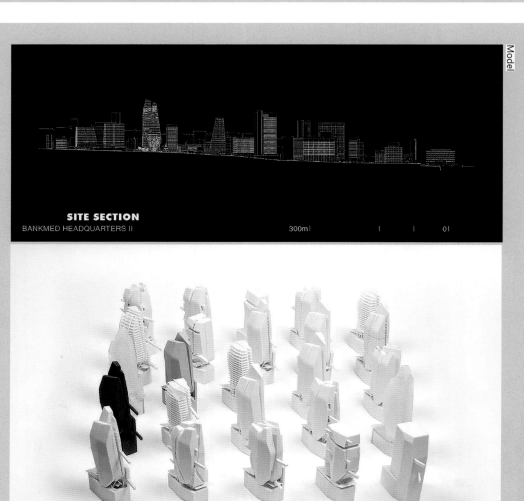

Site Section Plan / Model

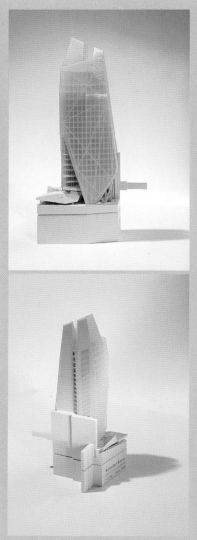

Model

Project Overview / 项目概况

City context

By virtue of its centrality and commanding historic setting, Beirut has accumulated a unique urban complexity that accounts for its survival as a comparatively open, pluralistic, and cosmopolitan community. The diversity of the urban environment also speaks of the city's tumultuous history, where a dynamic and at times violent intersection of political, economic and social forces has created a city polarised in many different enclaves.

1. The new BankMed HQ II is designed as two flutes, independent from each other while interconnected with each other, which indicates that the building will become the hub of the existing district and future development areas.

2. The naturally advantageous location, corner features, axis, positioning and unique building shape lay solid foundation for the building image.

3. The "double-flute" design divides the building into two parts, with one accommodating various functional amenities and services, and the other offices, public space and retailing banks.

4. The double-tower building is equipped with a case system, including the double elevation on single-layer glass of Tower East and spandrel glass of Tower West, to realise an overall goal of energy-saving.

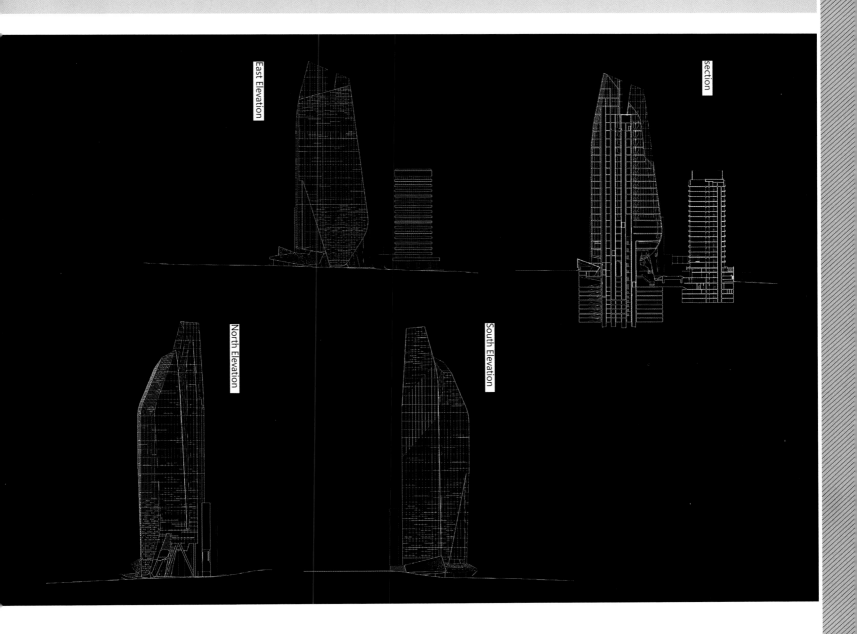

城市文脉
由于贝鲁特的集中性和其重要的历史背景，这里形成了独特而复杂的城市环境，导致其现存的群体十分开放、多元化和具有世界性。城市环境的多样化也显示在其动荡的历史上，政治、经济和社会的各派势力不断变化甚至时而暴力地交汇在一起，造成了这座城市在许多飞地（城市聚居区）上的两极分化。

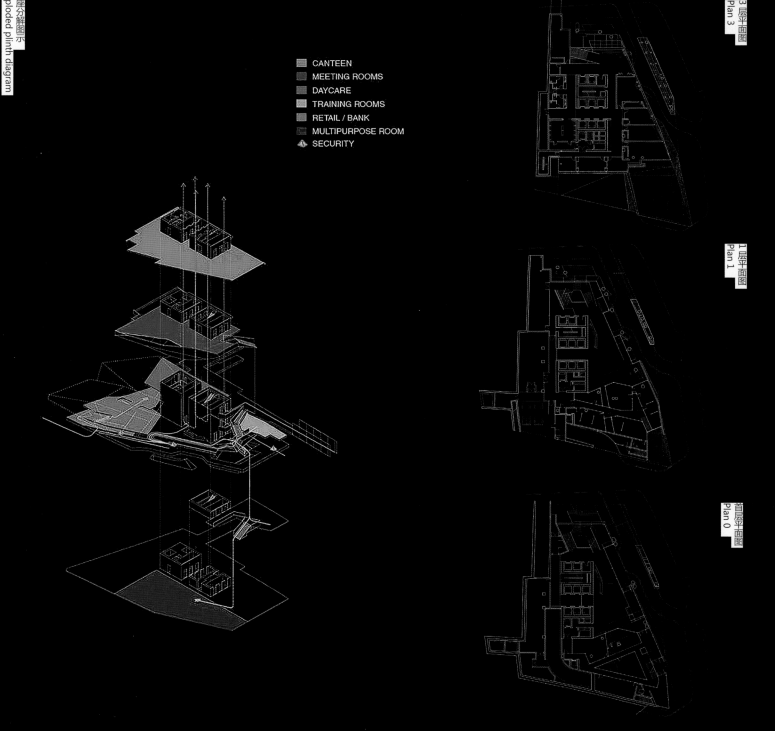

Upper level planning
In the current moment of growth, optimism, and rebuilding, new opportunities for rediscovering and rewriting the meaning of Beirut's urban richness present themselves. As a spirit of regeneration gains momentum, the need for new urban landmarks uniting Beirut becomes apparent.

Location analysis
Located outside the edge of Solidere development, the BankMed HQ II pushes a new frontier in Beirut's downtown growth, becoming an urban hinge between existing neighbourhoods and future developments. The site's natural vantage point, corner condition, axis, and positioning within a typological shift sets the stage for a unique formal response to the future needs of BankMed.

Design concept
Conceived as two vertical flutes, the form of the new BankMed HQ II is cleft to produce a slender tower with a striking presence on the skyline. The "two-flutes" concept allows the tower program to be clearly organised into halves, with one half containing functions and services, and the other containing offices, public spaces, and a retail bank. The building's northern aspect slopes to acknowledge BankMed's current offices in HQ1, with circulation bridges linking users in the existing office spaces to amenities provided in the new towers.

上层规划
在目前这个乐观地处于发展中的重建时期，重新发掘和重写贝鲁特城市财富的机会出现了。在重建精神的带动下，创造将贝鲁特团结起来的新城市地标的需求出现了。

区位分析
BankMed HQ II 位于 Solidere 开发区边缘外，为贝鲁特的市中心发展划定了新的边界，成为了现存街区和未来开发区之间的城市枢纽。现场天然的有利地点、拐角条件、轴线和定位为有条不紊地满足 BankMed 未来的需求打下了基础。

Design strategies

The overall mass of the tower fractures as it approaches the ground plane, framing a light-filled entry that draws movement from the street into the triple-height lobby space. Forms from the building's exterior extend into the lobby, creating a fluid transition from inside to outside. In the interior, the tower's cleft form results in dramatic glass wedges that shape unique office spaces, where irregularity in the form encourages informal gathering and dynamic workspace layouts.

Contributing to the overall energy efficiency of the building, the dual towers' exterior is defined by a system of skins including a reflective double facade over single-glazing on the east tower, and spandrel-glazed facades for the west. The rippled facade aggregates at the top and bottom of the tower, reducing heat gain while breaking the homogeneous reflections of a typical mirrored facade system.

设计理念
新的 BankMed HQII 被构想为如同两支竖着的长笛，因此其外部看起来如同分离成细长的塔形伸向天空。"双笛"构想让大楼可以被明确的一分为二，一半用于容纳各种功能性的设施和服务，另一半容纳办公室、公共区域和零售银行。大楼向北面阶梯式的倾斜以表示 BankMed 目前的办公室在 HQ1，使用者可以通过连接桥从现有的办公区域前往提供各种设施的新楼。

设计策略
整个大楼一直到底层都是分开的，在街道直到三层高的大厅中形成了一个明亮的入口。建筑外部的形状一直延伸入大厅内，让内外间的过渡变得流畅自然。在大楼内部，楼体的分离创造了引人注目的楔形玻璃，分割出了唯一的办公区，那里不规则的布局鼓励非正式的聚会和动态的工作空间布局。

为了达到建筑整体节能的目的，双塔楼的内部有一套外壳系统，包括东塔单层玻璃上带反射的双立面，和西塔的玻璃拱肩立面。波纹状的立面在大楼顶部和底部聚合，在减少热能的同时打破了典型镜面立面系统均匀反射的情况。

PROBLEM / 问题

1. **IMAGE SHAPING**
2. **TERRITORIALITY**
3. **MULTI-FUNCTION**
4. **ENERGY-SAVING**

1. 形象塑造
2. 地域性
3. 多功能性
4. 节能性

Ethiopian Insurance Corporation
埃塞俄比亚保险公司

Sohne & Partner
works
Sohne & Partner
作品

Architect: Sohne & Partner
Location: Ethiopia

设计公司：Sohne & Partner
地点：埃塞俄比亚

DESIGN REQUIREMENTS
设 计 要 求

Conceptually speaking – with structural metaphoric relationship – the Insurance Company embracing life and property of its customers is depicted as embracing nature by placing the tower and the podium in a natural landscape.

从概念上讲——从建筑的寓意来看——保险公司将大楼和裙楼建在自然景观附近，拥抱着自然，正如它拥抱着客户的生命和财产一样。

① 大厦由两部分组成,它们先相互依偎,而后成为一体,象征埃塞俄比亚保险公司与其客户之间相互依靠和信赖的关系。

② 设计中,建筑周围环绕的绿化区灵感来自典型的埃塞俄比亚风光。

③ 大楼包括了一个现代购物中心、一个城市娱乐中心、一个设有若干逃生房间的大礼堂以及一个多功能厅。

④ 大楼的外立面设有永久性防晒设备,以减少空调的使用。

Analysis

Main pedestrian flow on 1st level

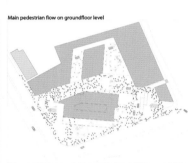
Main pedestrian flow on groundfloor level

Main office pedestrian flow on groundfloor

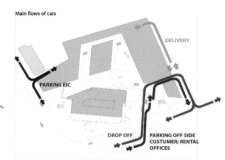
Main flows of cars

Structural Analysis

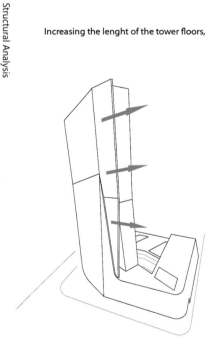
Increasing the lenght of the tower floors,

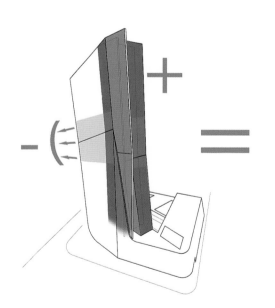
results in decreasing the hight

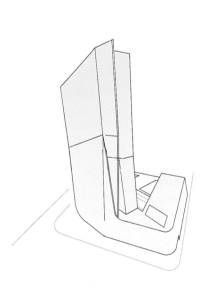
of the tower up to 10 levels

The building comprises two parts, leaning to each other, which indicates that the trust and reliance between Ethiopian Insurance Corporation and its customers.

The greened zones surrounding the building, flowing into and through the design, are inspired by the typical picturesque Ethiopian landscape.

The building consists of a modern shopping mall, an entertainment centre, an assembly hall with several small rooms for evacuation, and a multi-function hall. ❸

The outer wall of the building is equipped with permanently sunscreen facility to reduce air-conditioner use. ❹

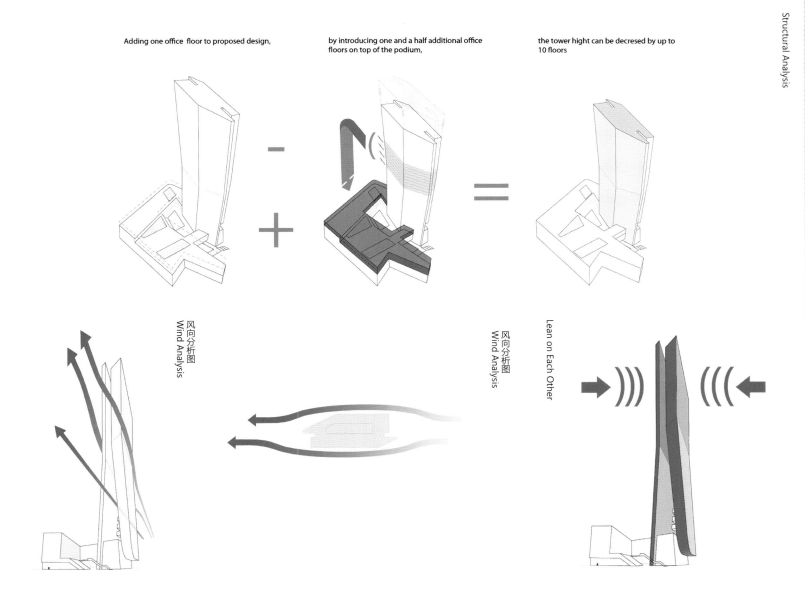

Adding one office floor to proposed design, by introducing one and a half additional office floors on top of the podium, the tower hight can be decresed by up to 10 floors

风向分析图 Wind Analysis

风向分析图 Wind Analysis

Lean on Each Other

Floors Analysis

Analysis

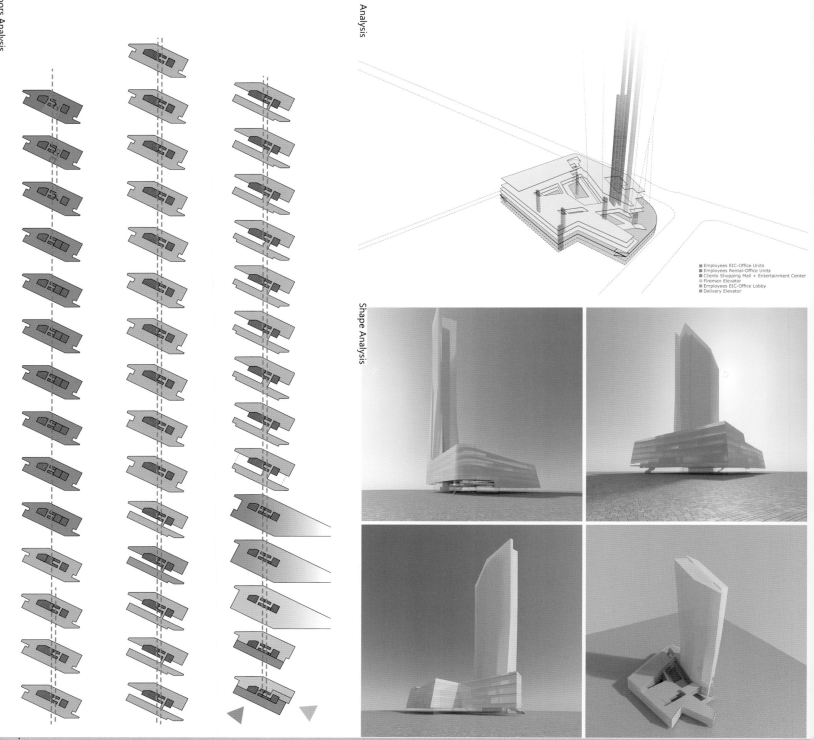

- Employees EIC-Office Units
- Employees Rental-Office Units
- Clients Shopping Mall + Entertainment Center
- Fireman Elevator
- Employees EIC-Office Lobby
- Delivery Elevator

Shape Analysis

Design strategies

The main recognition mark of the whole building complex is the Office Tower. It is made of two parts leaning on each other, becoming one unit at the end. It shows the connection between the EIC and its customers, depending and relying on each other. On the spacious plinth area of the tower a shopping and urban entertainment mall is located. The free space between the two parts of the tower is filled with green zones, terraces, bridges and loads of planted areas. It's another symbolic gesture of security, an image of a safe space provided from the Insurance Company for their customers. Instead of an antenna on top of the building the façades run together, forming a huge spire. Up there is a huge screen which is able to show moving pictures to be seen all over Addis Ababa.

设计策略

整个综合体的识别标志就是这座办公大厦。大厦由两部分组成，它们先相互依偎，而后成为一体。这体现了埃塞俄比亚保险公司与其客户之间相互依靠和信赖的关系。在大楼宽阔的基座区域设有一个购物中心和都市娱乐中心。大楼两部分之间的自由空间设有绿化带、露台、桥以及大量种植区域。这一符号象征着安全，代表着保险公司为客户所提供的安全空间。大楼的顶部没有装天线，而是让外立面交融在一起，形成了一个巨大的尖顶。顶上有一个巨大的显示屏，可以播放整个亚的斯亚贝巴都能看到的电影。

Function Analysis

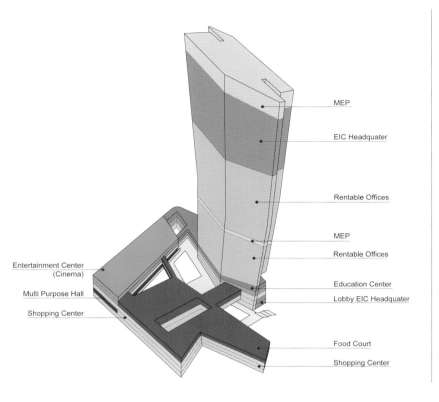
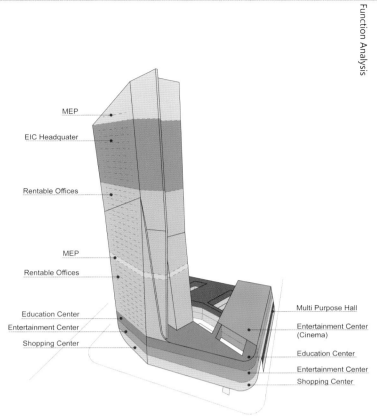

Shape Analysis

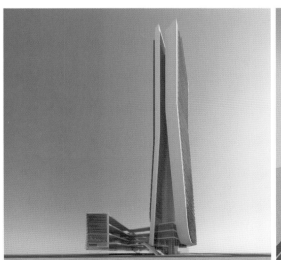

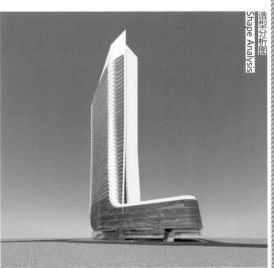

The EIC headquarters is located right on top of the tower, just where the two parts meet -- on the most powerful spot in the whole design. Besides the tower the design proposal also includes a modern shopping mall, an urban entertainment centre, an auditorium with several breakaway rooms and a multi-purpose hall, which is included in the design but is able to function totally on its own. A permanent external sun protection is included in the outside facade. This way the design helps reduce the usage of air conditioning systems and adds to the building's sustainability.

EIC总部设在大楼的最高层，正是两部分衔接的地方——是整个设计中最强有力的一笔。除了大楼之外，设计方案中还包括了一个现代购物中心、一个城市娱乐中心、一个设有若干递生房间的大礼堂，以及一个多功能厅。多功能厅包含在设计中，但是完全可以自行运营。大楼的外立面设有永久性防晒设备。这样的设计可以帮助减少空调的使用，同时延长了大楼的使用年限。

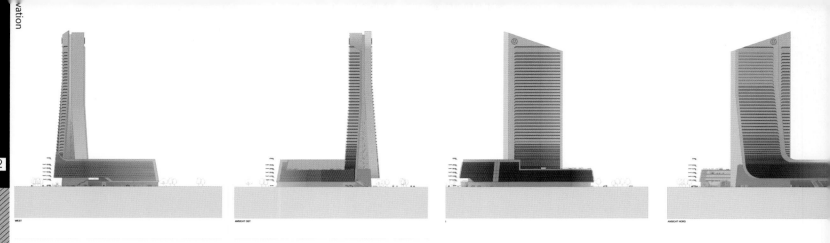

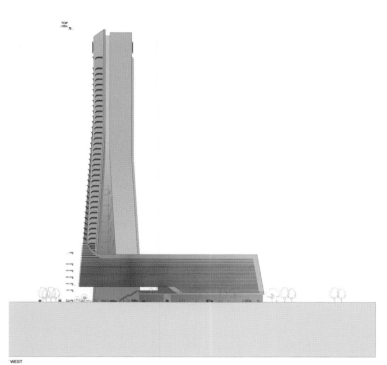

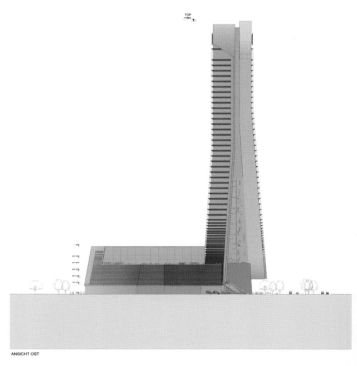

Sunlight Analysis

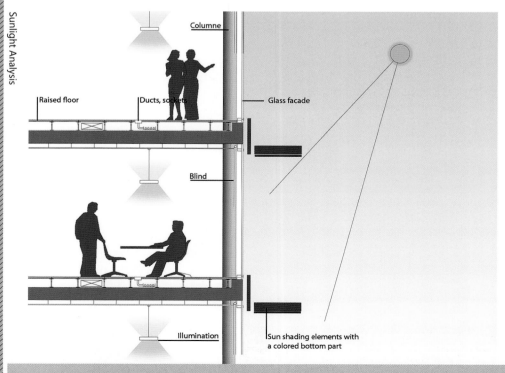

- Columne
- Raised floor
- Ducts, sockets
- Glass facade
- Blind
- Illumination
- Sun shading elements with a colored bottom part

Section

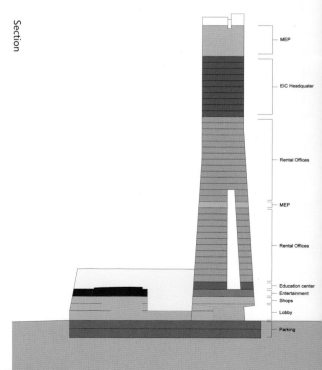

- MEP
- EIC Headquater
- Rental Offices
- MEP
- Rental Offices
- Education center
- Entertainment
- Shops
- Lobby
- Parking

PROBLEM / 问题

1. **NOISY SURROUNDINGS**
2. **CREATIVE WORKPLACE**
3. **HARMONY**
4. **UNIQUENESS**

1. 周边环境嘈杂
2. 创意工作空间
3. 回应场地
4. 独特性

100PP
100pp

MINISTRY OF DESIGN
works
MINISTRY OF DESIGN
作品

Architect: Ministry of Design
Client: CEL Development
Location: Singapore
Gross Floor Area: 12,600 m²
Function: Commercial building

设计公司：MINISTRY OF DESIGN
客户：CEL Development
地点：新加坡
建筑总面积：12 600 平方米
功能：商业建筑

DESIGN REQUIREMENTS
设 计 要 求

The new building is expected to re-define the features of commercial buildings and add values to the development project.

旨在通过新的建筑形式来重新定义商业建筑物的特性,并能为建筑开发项目增加可观的溢价。

THE SOLUTION / 解决方式

① 对大楼进行横向移位，创造一种像远离楼前的繁忙高架公路和前方的建筑的感觉。

② 项目模糊了商业空间和固定的功能空间之间的界限，为 21 世纪创造性的工作环境提供了激动人心的选择。

③ 在不同楼层引入一系列"阶梯式"阳台，贯穿不同的楼层，来回应场地的临海环境。

④ 大楼立面采用一系列横条分层，有意模糊每个构件的界线，产生横向视觉运动的感觉，强调了不同体量的位移和堆叠属性大楼，以产生了独特的天际轮廓线。

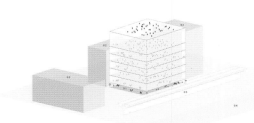

01-04 SITE CONSTRAINTS
TO MAXIMISE BUILDING MASS /
HIGHWAY OBSTRUCTS VIEW
最大可能的体量/高架阻挡视野
01 SITE 场地
02 EXISTING BUILDINGS 现状建筑
03 WEST COAST HIGHWAY 西部海岸路
04 PORT 海港

↑ LIFTED BUILDING MASS
TO ENABLE GROUND FLOOR POROSITY
抬起建筑
允许底层开放空间

···· DISTINCT FLOOR DIVISION
3M HEIGHT PER FLOOR WITH TOP 2 FLOORS COMBINED
3M层高且顶两层合并空间

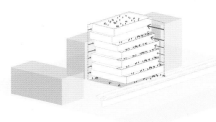

SHIFTED FLOOR MASS   FOR DISTINCT FLOOR DISTINCTION

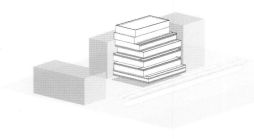

LAYERING (1)   INTRODUCTION OF AIR-CON LEDGES

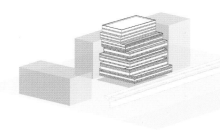

LAYERING (2)   INTRODUCTION OF STRIP WINDOWS

Project Overview / 项目概况

Ministry of Design's design for such a commercial building capitalises on these key traits and also introduces a number of key architectural gestures. These gestures aim to redefine the nature of such commercial buildings and also to provide an experience that adds a substantial premium to the development.

Firstly, to exploit the sea-fronting context of the site, the architects have introduced a series of "stepped" balconies across the different floor levels. These allow the building to appear to be shifting away from the busy elevated highway fronting the building.

The building is horizontally moved to create a sense of keeping away from the busy overhead highway and buildings in front of it.

The project blurs the boundary between commercial space and fixed function space, offering more exciting choices for creative work place.

A series of "stepping terraces" are introduced into different floors, to keep it in consistence with the environment near to sea.

The building's elevation adopts a series of horizontal layers to blur the boundary between each structural part, creating a sense of horizontal visual moving. It emphasises the movement of different bodies and stacked building, to create a unique skyline.

Concept

LAYERING (3)   INTRODUCTION OF COLOURED BANDS

Ground Floor Plan

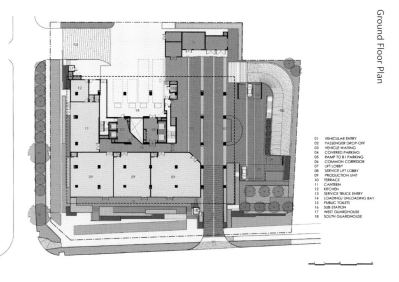

01 VEHICULAR ENTRY
02 PASSENGER DROP-OFF
03 VEHICLE WAITING
04 COVERED PARKING
05 RAMP TO B1 PARKING
06 COMMON CORRIDOR
07 LIFT LOBBY
08 SERVICE LIFT LOBBY
09 PRODUCTION UNIT
10 TERRACE
11 CANTEEN
12 KITCHEN
13 SERVICE TRUCK ENTRY
14 LOADING/ UNLOADING BAY
15 PUBLIC TOILETS
16 SUB-STATION
17 WEST GUARDHOUSE
18 SOUTH GUARDHOUSE

Typical Floor Plan

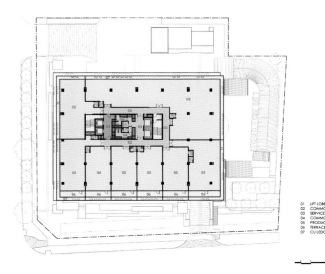

01 LIFT LOBBY
02 COMMON CORRIDOR
03 SERVICE LIFT LOBBY
04 COMMON TOILETS
05 PRODUCTION UNITS
06 TERRACE
07 CU LEDGE

Roof Floor Plan

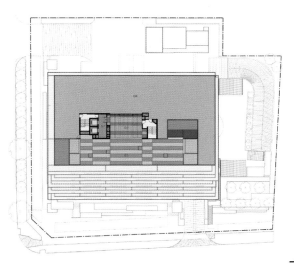

01 LIFT LOBBY
02 ROOF TERRACE W/ TRELLIS
03 ROOF TERRACE
04 MEP PLATFORM

设计理念
Ministry of Design 建筑设计公司通过充分利用这些关键性特点并引进各种关键性的建筑形式，成功设计完成了该项目商业建筑物。采用这些建筑形式的目的在于重新定义该项目商业建筑物的特性，并为该建筑开发项目增加了可观的溢价。

Secondly, the architects have also shifted the building laterally to create a sense that it comprises a series of dynamic blocks stacked one above the other rather than a static singular block. This allows the building to create a unique profile against the skyline.

Thirdly, the facade of the building comprises a number of different elements which are bound together aesthetically: primarily the windows, balconies and air-condition ledges. The architects have intentionally blurred the definition of each element by layering a series of horizontal stripes throughout the facade. The stripes generate visually movement horizontally across the building and also emphasise the shifting and stacked nature of the different volumes. A palette of varying greys is employed to generate the variety of tones required for the horizontal banding. This horizontal striping is also applied consistently to the landscape and hardscape elements surrounding the building.

设计策略

首先,为了充分开发出该项目地块的海岸背景优势,我们已经在建筑物的不同楼层之间设计了一系列"呈阶梯状"分布的阳台。如此一来,整栋建筑在视觉上就避开了位于正前方繁忙而嘈杂的高架公路。

其次,我们同时也对建筑物的侧面轮廓作了一些修改,从而营造出一种"建筑物是由一系列错落堆叠的动态模块组成"的即视感,而并非是静态的单一模块。这样一来,在城市天际线的映衬下,整栋建筑物的侧面轮廓独一无二。再次,

Lastly, the interior experience celebrates a stylised industrial aesthetic through the bold use of feature lighting, materials and environmental graphics across the different floors. Key interior spaces include the lift lobbies and passenger drop-off point and the building also provides a roof top garden space overlooking the sea.

When experienced in totality, the project blurs the boundaries between the predictable commercial space and the gritty industrial space, creating instead a hybrid space, which offers an exciting alternative for the creative workplace in the 21st century.

建筑物的外墙采用了一系列不同元素构造而成，经过我们巧妙的组合设计，颇具美感：主要体现在窗户、阳台和空调壁架设计上。通过在整面外墙上加上一层水平条纹设计，我们特意模糊了各种元素之间的界限。远望过去，外墙上的条纹仿佛会沿着水平方向波状起伏，更加突出了不同建筑体量之间错落平移的堆叠感。通过运用各种灰色系颜料，我们调配出适合装饰外墙横条纹的多种色调。与此同时，该项目建筑物周围的景观和硬景观也采用了与外墙一致的横条纹设计。

最后，通过在建筑物不同楼层之间大胆采用特色照明设计、材料和环境标识设计，我们在建筑物的内部装修中营造出了一种与众不同的工业美感。建筑物主要的内部空间包括电梯大堂、下客区等。同时，建筑物也设置了一个可俯瞰无敌海景的屋顶花园空间。

在整体上，该项目的设计模糊了可预见商业空间和砂质工业空间之间的界限，从而为 21 世纪创意工作空间提供一个令人兴奋的替代场所，而不是一个混合空间。

PROBLEM / 问题

1. **AESTHETICS VS. PRACTICABILITY**
2. **DAYLIGHTING & SUNSHADING**
3. **LIMITED SITE AREA**
4. **DETAIL OF GRADIENT FACADE**

1. 美学与实用平衡统一
2. 采光与遮阳兼顾
3. 有限的基地面积
4. 渐变立面的细部

Taiyuan Tower
太原中铁三局科技大厦

Henn
works
Henn
作品

Architect: Henn
Client: China Railway Engineering Group
Location: Taiyuan City, Shanxi Province, China
Site Area: 60,000 m²
Function: Office

设计公司：Henn
客户：China Railway Engineering Group
地点：中国山西省太原市
建筑面积：60 000 平方米
功能：办公

DESIGN REQUIREMENTS
设 计 要 求

The project is seated on the road cross of main roads in Taiyuan. Due to such special location, it needs to take into consideration the street landscape around. The style should be simple but modern, in consistence with the temperament of China Railway No.3 Engineering Group. The building covers about 60,000 square metres, in 200m height.

① 根据面积需要，建筑以能满足使用面积的最基本的几何形式-矩形为基础，结合使用功能和城市空间对建筑形体的影响，在面向十字路口对建筑平面进行退让处理，创造出良好的景观视野，也为城市创造出开阔的公共空间。通过二维平面的变化影响三维空间，建筑体量在首层为八边形，向顶部逐渐变化为倒角矩形，建筑的形体变化产生丰富的立面效果，形成城市雕塑般的建筑外观

② 建筑功能为办公。外立面采用宽幅玻璃幕墙使办公环境享有最舒适自然采光；竖向铝合金装饰造型同时起到遮阳作用。

③ 设计采用下小上大的体型，结构向上逐渐悬挑，在有限的基地面积上达到建筑面积最大化，并且使建筑看起来更加挺拔。

④ 立面造型为渐变的曲线。为实现这样的造型，建筑细部做法经过反复推敲，并结合消防排烟开启，保证实用功能不受影响。

Shape Analysis

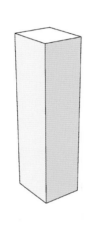

GENERIC MASS
基本体量

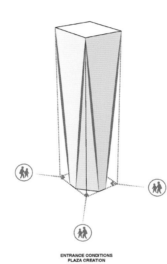

**ENTRANCE CONDITIONS
PLAZA CREATION**
建筑体退处理形成入口条件
创造入口广场

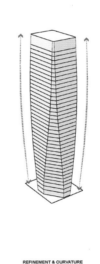

REFINEMENT & CURVATURE
重新定义形成曲线形体

Elevation Analysis

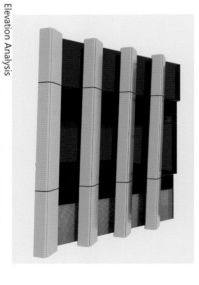

renders for reference only

Project Overview / 项目概况

Location analysis
This 200m high-rise in Taiyuan, Shanxi is located in a prominent position on the east-west axis of the central Chinese metropolis at the crossroad of Yingze Street and South Xinjian Road. The main function of this 60,000sqm building is the R&D centre of China Railway No. 3 Engineering Group. Main building 200 metres high, building density 25%, FAR 6.1. Two levels of basement for parking; above-ground area of the building are offices and conference rooms.

The architects took the most classic geometric form – "rectangle" – to meet the area demand. After considered the influence of the function and urban space to the shape of the building, a "setback" has been used. As a result, the architects created a great view for the building ane also a wild public space for the city. The building volume of the first floor is octagon, gradually changed to chamfered rectangular at the top of the tower. The changing of the building appearance formed a varied facade. It created a practical tower with aesthetics of sculpture artwork.

Main programme of the tower is office. Large size glass facade provides office area with a comfortable environment with enough daylight; at the same time, the vertical facade elements play a key role as sun shading.

Building volume grows bottom up in order to have a max. building area at a very limited site. And it also makes the building looks tall and straight.

Gradient curve facade elements require smart details, which must also fulfill the fireproof requirements like smoke exhaust.

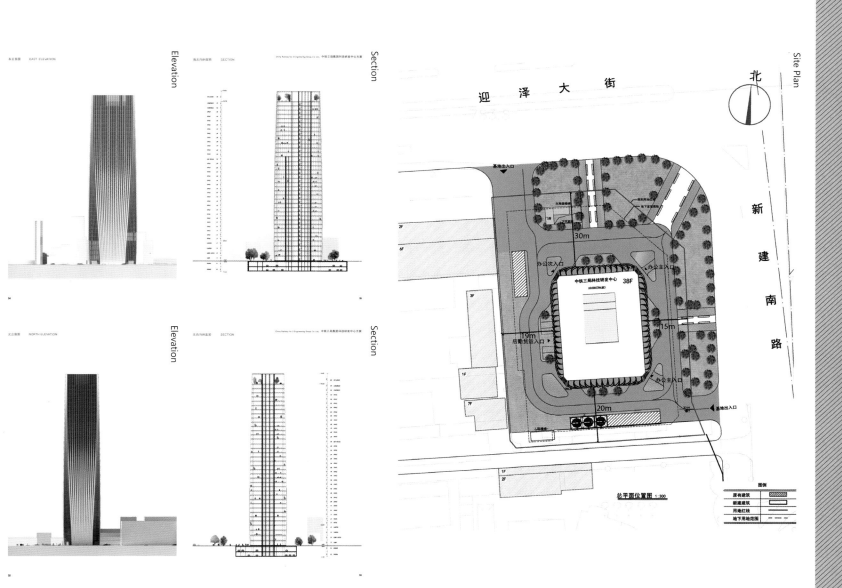

区位分析

由 HENN 建筑设计的 200 米超高层项目位于山西省太原市的城市东西轴线的突出位置，迎泽大街与新建南路交汇口，建筑面积约 6 万平米，为中铁三局办公及研究中心。主楼高 200 米，建筑密度 25%，容积率 6.1，地下两层为停车，地上为办公及会议用房。

Facade Concept

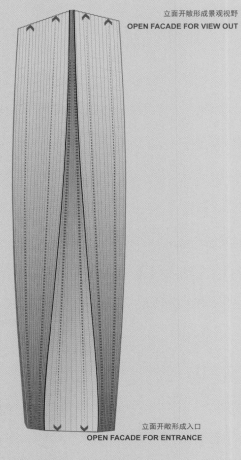

立面开敞形成景观视野
OPEN FACADE FOR VIEW OUT

立面开敞形成入口
OPEN FACADE FOR ENTRANCE

立面百叶密度变化创造建筑入口条件
DENSITY CREATES ENTRANCE CONDITION

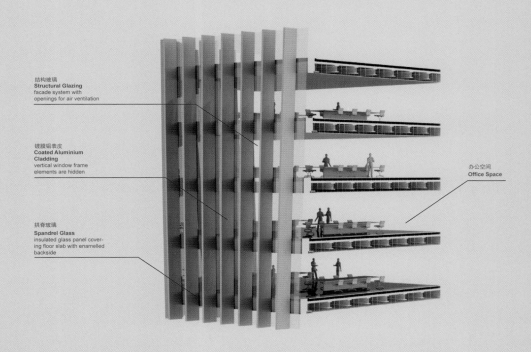

结构玻璃
Structural Glazing
facade system with openings for air ventilation

镀膜铝表皮
Coated Aluminium Cladding
vertical window frame elements are hidden

拱脊玻璃
Spandrel Glass
insulated glass panel covering floor slab with enamelled backside

办公空间
Office Space

1st Floor Plan
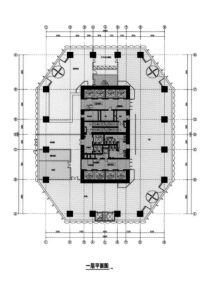
一层平面图

2nd Floor Plan
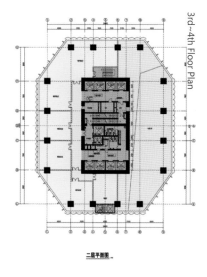
二层平面图

3rd~4th Floor Plan
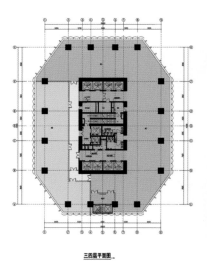
三四层平面图

Design strateies

The design of the 200m high tower makes optimum use of the site and is based on a traditional geometric volume, which starts from an octagonal footprint and transforms dynamically to assume a generally rectangular appearance. This results in clearly accentuated entrances so that the lower storeys of the building look out onto the city and its corner position allows ease of access from all sides.

设计策略

200米高的塔楼设计基于传统的几何形状，从八角形足迹开始，将塔尖转变为一般的矩形外观。这导致入口明显加重，在建筑物的下面几层可以眺望这座城市，其角落位置易于从各个方向进入。

入口区域的拐角完全是玻璃的。立面表面的线条和元素密度符合建筑几何的逻辑，表现了从实心到开放表面的温和转换。这种纵横交错的图案最终在较高的楼层达到最大的玻璃比例，允许自然光进入。

The chamfered corners of the entrance areas are fully glazed while the profiling of the remaining facade surfaces in terms of their lines and density of elements follows from the logic of the building geometry and so describes a gentle transformation from solid to open surfaces. This criss-cross pattern finally opens to reach its maximum proportion of glass in the upper storeys and allows a high level of natural light into each floor while drawing the eye to the sculptural qualities of this building.

PROBLEM / 问题

1) CREATE PROJECT WITH A LOW OR ZERO CO_2 CARBON FOOTPRINT

2) MAXIMISES OUTLOOKS, SEA VIEWS AND SUNLIGHT QUALITIES

3) MAXIMISE SPATIAL QUALITIES

1. 创建一个低碳或零排放项目
2. 视野开阔、海景可观、采光良好
3. 空间质量上乘

Wooden Highrise Apartments
RIVM & CBG 总部大楼

Tham & Videgard Arkitekter
works
Tham & Videard Arkitekter
作品

Architect: Tham & Videgard Arkitekter
Client: VolkerWessels Bouw & Vastgoedontwikkeling
Location: Utrecht, The Netherlands
Building Area: ca. 21,000 m²
Function: Research institutes

建筑师：UNStudio
客户：VolkerWessels Bouw & Vastgoedontwikkeling
地点：荷兰乌得勒支
建筑面积：约 21 000 平方米
功能：研究机构

DESIGN REQUIREMENTS
设 计 要 求

Stockholm develops vigorously, while Loudden is expected to develop into a new city district. Therefore, this new development project requires the overall goal — working out a truly sustainable solution for the long run.

① 从承重结构到建筑正面，从饰面到门窗，整个建筑仅采用一种材料——瑞典实木。瑞典森林能够快速满足建筑所需木材需求。

② 建筑塔楼之间的间隙较大，住户可由此观看海景。同样地，利用这种设计，日光亦能够直接照射到北向的码头通道。

③ 设计引用城市街区概念，包含一些低层办公区和商铺和四座细长塔楼。结构打破了海边码头标准的卡雷模式，更能充分利用场地空间。塔楼和三层基座仍构成城市街区的主要部分。这种折叠式平面形状沿着一个既挡风又有明媚阳光的"双排"巷子形状，为新城市结构创造各种开放式会议空间和室内活动空间。

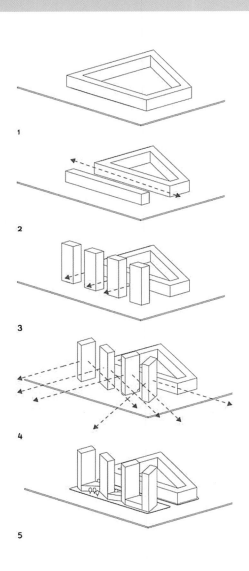

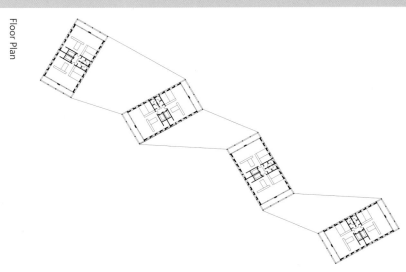

Upper level planning
Stockholm is growing fast and the Loudden area, the former international transport harbour, on the northeastern island of Djurgården, is planned to be converted into a new urban area. In line with the overall direction for this new development to achieve real long term environmentally sustainable solutions, Tham & Videgård Arkitekter have developed a proposal for apartment blocks of solid wood.

Feasibility analysis
The project investigates the possibility of building high rise buildings along the quay next to a planned six-storey urban block structure following the classical carré typology. The project thus aims to create new urban and spatial qualities through careful study of placement, outlooks, solar angles, street section and architectural form. The selected site in the northern part of the former harbour area sits directly on the quay next to the sea.

上层规划
斯德哥尔摩的发展势态正猛，而Loudden地区（之前是一个国际运输港口，位于东北方向的Djurgården岛屿上）的发展计划是转型为一个新的城市地区。根据该新型开发项目的总体发展目标——制定真正的长期环境可持续发展解决方案，Tham & Videgård Arkitekter已经提出建设实木结构公寓楼的设计方案。

可行性分析
该项目设计团队已经调查了沿着码头建造高层住宅楼的可行性，而紧挨着该项目住宅楼的是一栋根据传统建筑类型学计划建造的六层高的城市建筑物。在对项目方位、外观、阳光照射角度、街区以及建筑形式作出仔细研究的基础上，最终确定该项目的目标是创建一栋空间布局特点鲜明的新型城市大楼。为该项目选择的地块位于前港口区的北面，恰好在海边码头之上。

Project Overview / 项目概况

Design strategies
Four 20-storey apartment buildings form together a new landmark, a cornerstone within the new area. Gaps between the buildings leave open views from the block behind towards the sea and also let direct sunlight reach to the north-facing quay promenade.

设计策略
四栋20层高的公寓大楼将一起成为该新区占据重要位置的地标式建筑。每两栋公寓楼之间相隔的距离使得从后方街区欣赏海景的视野非常开阔，而且阳光可直接照射到朝北的码头人行漫步道。各栋高层塔楼共享一栋三层高的基楼，从而形成清晰的街区布局。该项目设计的折叠围合式室外空间不仅挡风，而且日照充足，非常适宜举办会议和一些户外活动。从框架到外墙，再到表面装饰和门窗，所有四栋建筑大楼都只使用同一种材料，即瑞典实木。通过使用一致性的可再生材料，比如木材，该项目建筑大楼将符合可持续发展的要求，并且隔热性能非常好，建筑结构也非常稳定牢固。随着时间的推移，使用木材的优点将越发突出，而且能最小化该项目建筑大楼的总体能源消耗。较低层基楼的屋顶上将种植满景天属植物，以充分利用当地的雨水条件。但是，四栋公寓楼的屋顶将安装太阳能设备。在每一栋公寓楼的顶部，都设有一个可用于举办各种娱乐和社交活动的共享冬季花园。

The buildings are constructed entirely in one material, Swedish solid wood, from the load-bearing structure to the facade, finishes and windows. The Swedish forest produces the acquired amount of wood in five minutes.

The gaps between the housing towers allow open views towards the sea for its residences, along with the block behind. In the same manner the scheme also let direct sunlight reach the north-facing quay promenade.

The scheme introduces an urban block that consists of a low base for offices and shops, and four slender housing towers. The structure breaks the sequence of the standard carré typology in order to take full advantage of the site, at the quay right next to the sea. The combination of towers with a three-storey base still supports a clear urban street section. Its folded plan shape creates a series of exterior places for meeting and outdoor activities along a "second row" alley in a wind protected and sunny position within the new urban structure.

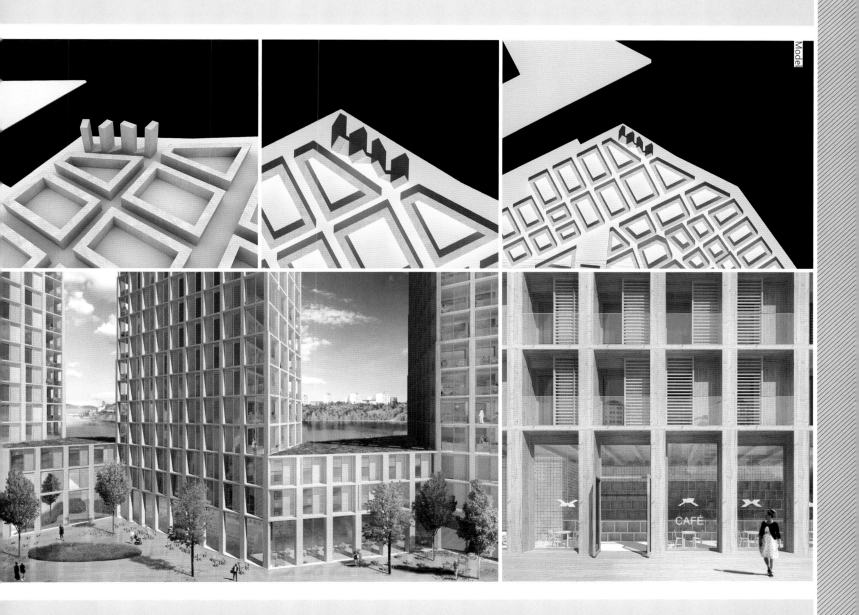

The high rise towers are interconnected by a three-storey base that supports a clear street section. Its folded plan shapes exterior spaces for meetings and outdoor activities in wind sheltered and sunny locations. The buildings are constructed entirely in one material, Swedish solid wood, from the frame to the facade, finishes and windows. Through consistent use of a renewable material like wood, the result is a sustainable, well-insulated and robust house structure with good potential to perform well over time, and minimise the total energy consumption.

The roof of the lower base will be covered with sedum plants that take care of rainwater, while the roof of the four towers will be fitted with solar cells. At the top of each tower, there is a common winter garden for recreation and social activities.

PROBLEM / 问题

1. **MULTI-FUNCTION**
2. **ADAPTABILITY TO SURROUNDING ENVIRONMENT**
3. **UNIQUE FEATURES**
4. **ECONOMICAL DRIVE**

1. 功能多样性
2. 适应周边环境
3. 独特的特性
4. 促进当地的经济

New Keelung Harbour And Service
中国台湾新基隆港服务大楼

MSA
works
MSA
作品

Architect: MSA
Principal in Charge: Maxi Spina
Design Team: Cesar Beltran, Paul Castellanos, Miriam Jacobsen, Alan Manning, Kendra Ramirez, Richard Solis and Joseph Veliz.
Location: Keelung, Taiwan, China
Area: 82,000 m²
Function: Cruise ship terminal and port service centre

设计公司：MSA
建筑师：Maxi Spina（主要负责人）
设计团队：Cesar Beltran, Paul Castellanos, Miriam Jacobsen, Alan Manning, Kendra Ramirez, Richard Solis 和 Joseph Veliz.
地点：台湾基隆
面积：82000平方米
功能：邮轮码头与港口服务中心

造型分析图
Shape Analysis

1

CITY
ROAD
SEA

TYPICAL TRANSPORT NODE

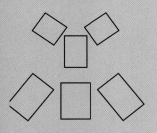

2

TYPICAL TRANSPORT - BUILDINGS DIVIDED INTO WINGS

1+2 = 3

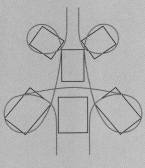

LINKED INDIVIDUAL MASSES

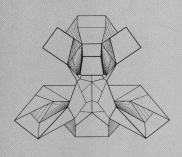

RESULT

DESIGN REQUIREMENTS
设 计 要 求

The international competition for the New Keelung Harbour Service Building held in 2012 asked for a new facility able to accommodate a dozen thousand cruise ship passengers a day plus a large number of administrative functions of the Port of Keelung.

在 2012 年举办的新基隆港服务大厦国际设计竞赛中提出，需创建新型设施，每天可容纳上万巡游船游客，以及大量基隆港行政管理功能设施。

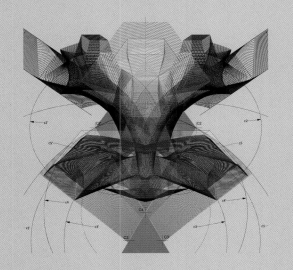

CONTORTED SHELL SHOWING FRACTURED PATTERN

COMPLETE BUILDING SHELL

PLANAR SHELL SHOWING REGULAR PATTERN

THE SOLUTION / 解决方式

① 借用流动性、蝴蝶结式连接性的设计手法以及清晰利落、基于海湾的或者基于侧翼的空间组织布局方法将整栋项目大楼的建设分成三个部分。保持该大楼服务功能、旅游功能、休闲娱乐功能以及商业功能之间的连贯性，同时也保留诸多功能之间的区别。

② 在形式上，该项目大楼采用了扭曲式设计的手法，通过一系列有规律的退台设计，悬臂式设计以及部分延伸设计，巧妙地将周边的建筑物连接在一起，从而形成一种无缝式的空间体验。

③ 项目大楼的外观将随着整栋大楼的扭曲与折叠变化而变换其图案、配色以及孔洞几何构造，从而在海港服务站大楼的终点与连接性区域形成变化无穷的空间布局与光影效果。

④ 设计一条高架于项目地块之上的公共人行漫步道，用来分流旅客。将开放式的公共空间、商业、旅游以及休闲娱乐功能连接起来，不仅使这个公共区域充满活力，还给旅客们和城市居民带来一段独特的体验。

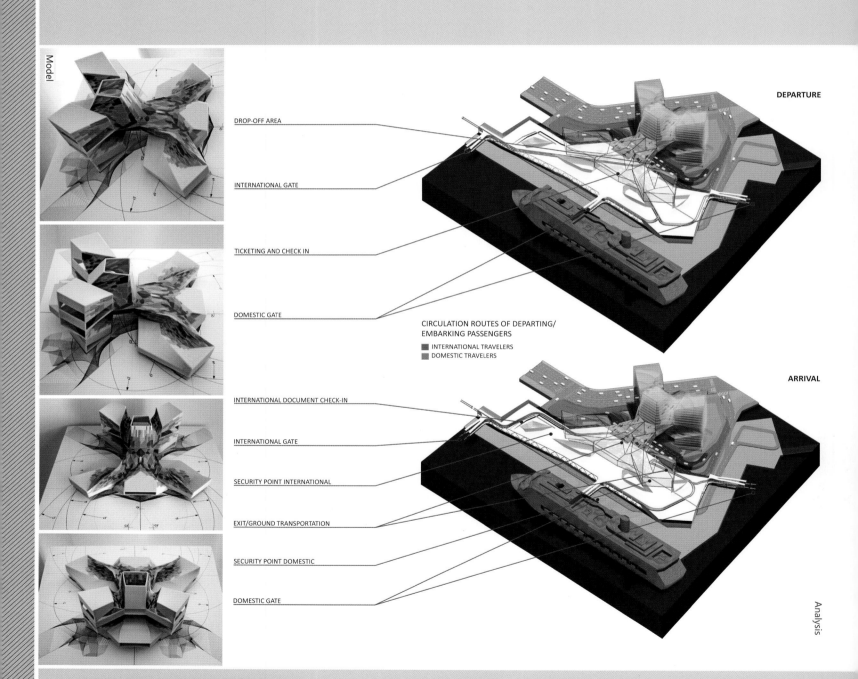

Project Overview / 项目概况

Location analysis

Through this state-of-the-art transportation interchange, the project articulates a Ferry and Cargo Terminal, an elevated Public Promenade in the sea side of the site, and the Harbour and Service Building and Commercial spaces to the Northeast. Infrastructure and service buildings have usually been characterised by the specificity and efficiency of their connections.

区位分析

作为目前技术水平最高的客流集散枢纽中心，该新基隆港服务大楼建设项目同时也包括一个轮渡与货运码头、一条靠近海边的高架公共人行慢步道、海港服务大楼以及位于东北面的各大商业空间。通常来说，基础设施和服务大楼主要是

The whole building is divided into three parts via flowing or bowknot connection, and clear, harbor-based or wing-based space layout, to ensure the consistence among the building's service, tourism, resort and commercial functions while keeping their respective features.

In terms of shape, the building is perfectly connected with other buildings around by warping, set-backs, hanging wings and partial extending, providing a seamless space experience.

The building changes its patterns, colours and hole structures externally with its curving and folding differences, creating a varying space layout and light effect between the terminal and connecting area of the harbour and service building.

A public sidewalk is designed over the project block for the passengers, connecting with the open public space, commerce, tourism and entertainment functions. It gives more vitality to the public area, and brings about a unique experience for tourists and urban residents as well.

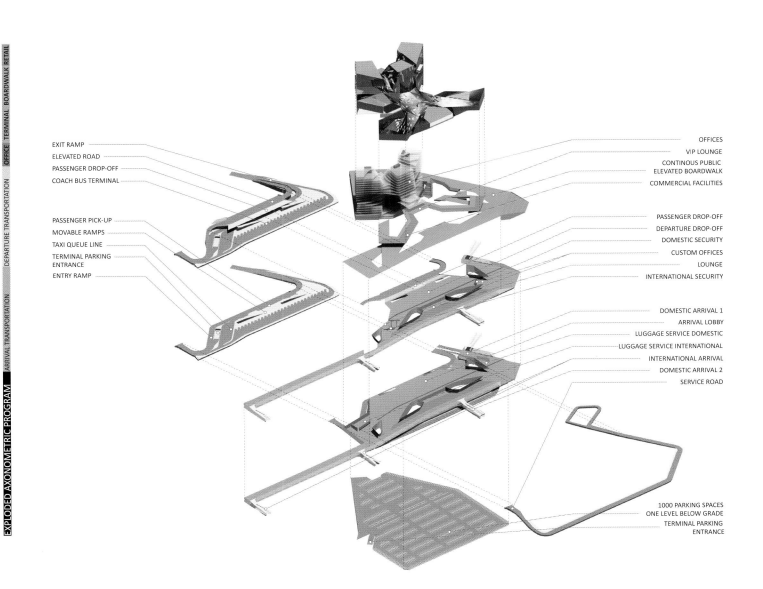

This paradigm has resulted in two distinct typologies: the fluid ones, borrowing from the dynamism and speed that characterises travel and movement in general; and the rigid clusters of buildings, creating strictly differentiated volumes arranged in fixed layouts composed of wings or bays.

以它们之间形成联系的特殊性和效能为主要特色的。该项目范例已经造成了两种具有明显不同的类型学分类：一种是流动式布局，借用动力机制和速度原理，一般以旅游和活动为主要特征；另一种则是建筑物的固定集群布局，将严格区分的各大体量安排设置在基于侧翼或基于海湾地形的固定布局中。

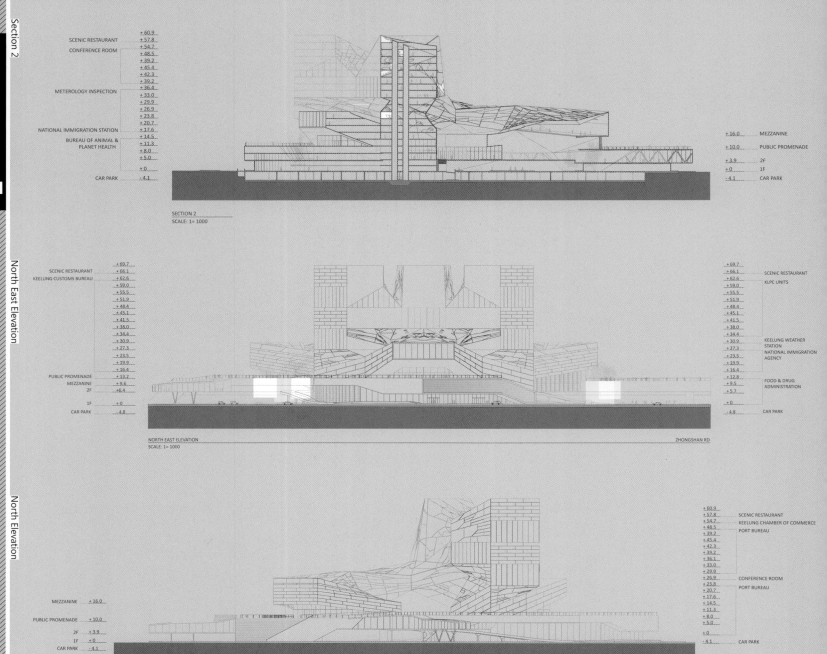

Invigorated City Edge at the Water

The passenger and cargo terminal as envisioned by MSA is designed to revitalise the Keelung urban shoreline through an elevated public promenade above the plinth that houses most of the passenger functions. This new urban space connects with a range of recreational, commercial and service spaces in the upper levels of the terminal, as well as with the new project development in the KURP area. Furthermore, this path from the promenade through the mezzanines continues into the Harbour and Service building and offers aerial views of the city and the hill towards the north. At the water edge, the elevated promenade culminates grand stair, which provides access to a maritime gateway plaza. This linkage of open public spaces, commerce, travel and leisure activates a vigorous public domain providing a unique experience to both travellers and city dwellers.

New State-of-the-Art Transportation Interchange

The proposal brings Zhongshan Rd to the heart of the New Harbour and Service Building by extending this urban avenue into the site as a multi-storey transportation hub, differentiating traffic flows into three levels: an underground level creates space for over 1,000 vehicles, while the ground floor allocates parking for 28 coach busses, taxi queue line and a dedicated lane for pick-up of arriving passengers. In its turn, the second floor distributes both departing traffic of the ferry terminal as well as dedicated traffic for the new Harbour and Service building. In this way, this central multi-avenue distributes traffic from Zhongshan Road into the site and then collects it into Lane # 36 Zhongshaner Rd, with frank exits for both North and Southbound traffic. In addition, this central avenue marks the division between Phase I and II of the project construction, allocating the Ferry and Cargo Terminal in the sea side of the site and the Harbour and Service Building and Ancillary Commercial space to the Northeast.

生机勃勃的滨海城市边缘区

由 MSA 构想的客运和货运大楼将被设计用来振兴基隆城市海岸沿线的发展，高架于该项目地块之上的公共人行漫步道将主要发挥其分流旅客的功能。这一新型城市空间与集中分布于运和货运大楼上半部分楼层的各种娱乐、商业和服务空间，以及 KURP 地区的新项目开发空间连接在一起。而且，从人行漫步道贯穿至夹楼的这一条路线一直延伸至海港服务大楼内部，是欣赏北面城市风景和山景的绝佳高空位置。在海岸边，高架人行漫步道与宏伟的阶梯相接，人们可以经由此处直接到达海上入口广场。通过将开放式的公共空间、商业、旅游以及休闲娱乐功能连接起来，不仅能使这个公共区域充满活力，而且能给旅客们和城市居民带来一段独特的体验。

Design concept

While these two models of space organisation have traditionally been in opposition, the architects recognise open spatial possibilities in service and transport buildings not yet coded with an architectural typology. Thus borrowing as much from fluid, knot-like, connective design as well as from clear, bay or wing-based space organisation. Along that line of thinking, MSA's design for the New Keelung Harbour and Service Building proposes a three-part based typology that subtly contorts its mass in order to link to neighbouring spaces and activities, achieving overt continuity between service, travelling, leisure and commerce, while at the same time maintaining differentiation among the many functions.

Design strategie
Contorted Form, Differentiated Spaces

Formally, the building inflects and contorts its mass in order to strategically connect with adjacent programmes through a series of controlled setbacks, cantilevers and overhangs, creating a seamless spatial experience; the three-part layout work as much for the Harbour and service building as for the passenger terminal. In the former, two 20-storey towers are located to the northeast side of the central hub and linked to a third tower, which hovers above the central road and connects to the centre volume of the terminal on the sea side; similarly, the passenger terminal is modulated into three main volumes, housing the lobby and ticketing area at the centre, and the domestic and international terminals at each side. Formally interconnected, yet programmatically and typologically differentiated, the design creates a unique hyper connected building that performs both as a cluster of functions and spaces as well as a single, exciting urban hub.

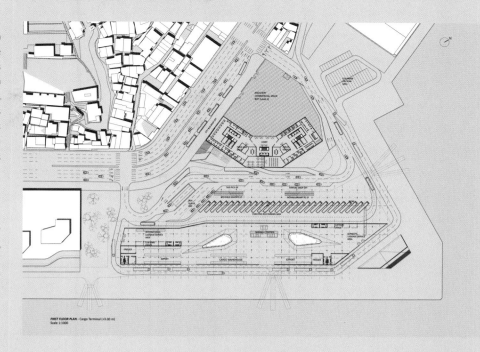

1st Floor Plan

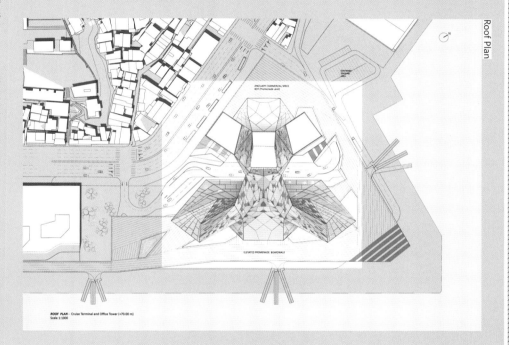

Roof Plan

设计理念

从传统意义上讲,这两种空间组织布局模式虽然是相互对立的,但是我们仍然认为服务和运输大楼的设计存在开放式空间布局的可能性,并不一定要遵从任何一种单一建筑类型学的原理。因此,我们决定借用流动性、蝴蝶结式连接性的设计手法以及清晰利落、基于海湾的或者基于侧翼的空间组织布局方法。按照这个思路,MSA公司对台湾新基隆港服务大楼的设计方案建议根据建筑类型学的原理,可将整栋项目大楼的建设分成三个部分。其中,该项目大楼体量部分将采用扭曲设计,巧妙地将周边的空间和活动连接起来。该设计方式不仅能够保持该大楼服务功能、旅游功能、休闲娱乐功能以及商业功能之间的连贯性,同时也能保留诸多功能之间的区别。

设计策略
扭曲式形式设计,有区别的空间

在形式上,该项目大楼采用了扭曲式设计的手法,通过一系列有规律的退台设计、悬臂式设计以及部分延伸设计,巧妙地将周边的建筑物连接在一起,从而形成一种无缝式的空间体验;不仅海港服务大楼采用了三分布局设计的方法,旅客航厦的设计也采用了这种布局方法。就海港服务大楼而言,两栋高达20层的塔楼位于中心枢纽站的东北侧,并且与第三栋塔楼连接在一起。第三栋塔楼从中心大道上盘旋而上,并与海边的旅客航厦的中心体量部分连接在一起;同样地,旅客航厦也被设计成由三大主要体量模块组成,中间的大楼主要包括一个大厅和售票区,而国内航厦以及国际航厦分别位于两侧。在形式上相互连接,然而在布局方法和类型上加以区别的设计手法将创造出一栋独特的大楼(连贯性超强),既集合了诸多功能和空间,也可作为一个独立而令人兴奋的城市枢纽中心投入使用。

Articulation, Fenestration and Pigmentation

A public building in such a prominent urban site necessitates a new approach to material form; one that is able to provide a new type of atmosphere to the spatial sequence that brings travellers, shoppers and city dwellers closer together. The project takes an active part on this proposition through a dynamic interplay between formal articulation, fenestration, and colouring. In this way, the building's shell changes its geometric constitution in terms of pattern, coloration and aperture as the mass of the building turns and folds. This creates ever-changing spatial and light effects within the terminal and the connective regions of the Harbour Building, while the brushed steel exterior shell periodically changes its colour from Turquoise, to Sea Green, to Emerald to Olive Green through the rhythms of a fracturing geometric motif. This integral approach to the material life of the building brings it closer to its vibrant civic life as well as to the rich and changing natural geographic conditions around it.

最先进的客流集散枢纽中心

根据我们的设计,中山路应位于新基隆港口服务站大楼的中心位置,并同时将城市大道扩展延伸至该大楼建设地块,以形成一个多层的交通枢纽中心,分三个楼层将交通流区别开来:地下一层的空间可容纳一千多辆汽车;地面一层可设置28个大客车停靠位、出租车排队区以及一条接送入境乘客的专用车道。照此顺序,大楼的第二层可用于分散轮渡码头离境的交通流以及用作新海港服务站大楼的专用交通道。如此一来,贯穿该地块中心的多条大道便可以帮助分散从中山路涌入该地块的交通流,然后再将交通流集中引导进中山二路的36号车道;同时往南北方向去的交通流可在侧边的出口通行。此外,这条中心大道也是该项目建设一期与二期工程之间的分界线,轮渡和货运码头站大楼位于地块靠海的一侧,而海港服务站大楼与附属的商业空间则坐落于东北侧。

巧妙连接,门窗设计以及色彩搭配

在这样一块处于显要位置的城市地块上建造一座公共大楼需要一种实现物质形态的新方法;这种方法应该是能够为项目的空间序列设计提供一种新型的风格,可增加游客、商户以及城市居民之间的互动和亲切感。通过各部分之间的巧妙连接、门窗布局和色彩搭配以及其动态的相互作用关系,我们设计的建筑项目在这方面发挥着积极作用。如此一来,该项目大楼的外观将随着整栋大楼的扭曲与折叠变化而变换其图案、配色以及孔洞几何构造,从而在该栋海港服务站大楼的终点与连接性区域形成变化无穷的空间布局与光影效果。但是,抛光钢制外墙的表皮将会周期性地变换其表面颜色,随着几何图案重组渐变的节奏,从天蓝色到海绿色,再变换成翠绿色,最后变成橄榄绿色。这种激活整栋大楼物质生命的完整方法将使该项目与居民们活力四射的日常生活联系得更为紧密,而且更能与该地丰富而不断变化的自然地理条件和环境融合于一体。

PROBLEM / 问题

1) ENVIRONMENT PROTECTION
2) COMPLEX FUNCTIONS
3) TERRITORY

1. 环境保护
2. 功能流线复杂
3. 地域性

Baltic Parnu
Baltic Parnu

SPARK
works
SPARK
作品

Architect: SPARK
Project Director: Mingyin Tan
Team: (Project Architect) Uldis Sedlovs, Yuchen Zuo, Adrian Garcia
Client: Competition entry – Rail Baltic Pärnu Passenger Terminal
Location: Parnu, Esthonia

设计公司：SPARK 思邦
项目总监：Mingyin Tan
团队成员：(Project Architect) Uldis Sedlovs, Yuchen Zuo, Adrian Garcia
客户：Competition entry – Rail Baltic Pärnu Passenger Terminal
地点：爱沙尼亚帕尔努

DESIGN REQUIREMENTS
设 计 要 求

The terminal of Baltic Parnu is in the south of Parnu town. After repeated transportation evaluation, the existing premise will become the hub for all public transportation.

Baltic Parnu 终点站位于 Parnu 河滨，Parnu 城镇的南面。经过反复的交通评估后，现有的场地将会作为未来所有公共交通的共同港湾。

THE SOLUTION / 解决方式

① 设计为延续空间以及其外部自然环境的沟通，在空间的处理方式上试图与周边环境建立强而有力的关系。

② 设计方案为通透的两层建筑，通过视觉开放性和纵向流线，保持地面公共空间与上层后勤空间的连贯性。

③ 设计中重点使用源于爱沙尼亚历史文化以及当地手艺人精湛建筑技艺的木质材料。

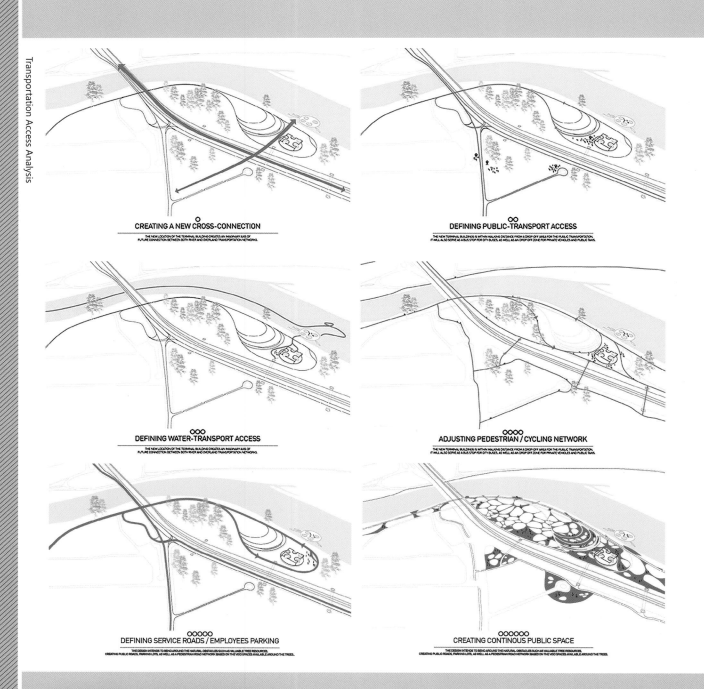

Transportation Access Analysis

Project Overview / 项目概况

Design concept

New Rail Baltic Parnu passenger terminal proposal has been designed to visualise a "seat" in the forest, a metaphysical space under the pine trees at the coast of the Baltic Sea. Inspired by the qualities of a piece of dressed wood, the piece of architecture tells a story about the tree itself, while the new terminal building compliments the contemporary architectural setting in order to establish a strong and reflexive relationship with nature and its intermediate surrounding environment. Designed as a place for sustained dialogues within the natural

设计理念

新的 Baltic Parnu 终点站的设计概念从一个森林中"憩席"出发，制造出类似 Baltic 海岸边松树下的抽象空间。设计的灵感源自于一块打磨过的松木，整体建筑讲述了一个树木本身的故事。不仅如此，设计还同时结合了现代建筑的背景去展现大自然与周边环境强而有力的关系。

In order to better the connection with natural environment, the space design is intended to establish a close relation with its surroundings.

A two-storey well-ventilated building is designed to keep continuous between the public space on the ground and logistics space on upper floor through visual and vertical streamlines.

The design values the use of wooden materials that present the Estonian culture and local artists' outstanding building skills.

○
DEFINING FUNCTIONAL LAYOUT

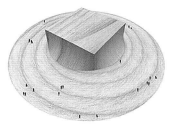
○○
TRANSFORMING VISUAL IDENTITY

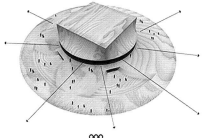
○○○
RELEAVING VIEWS & ACCESS

○○○○
PROVIDING NATURAL DAYLIGHT

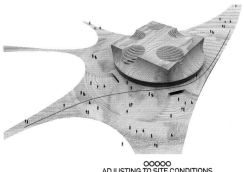
○○○○○
ADJUSTING TO SITE CONDITIONS

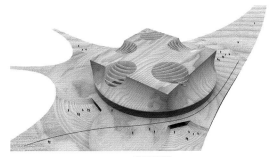
○○○○○○
PUBLIC SPACE BENEATH PUBLIC SERVICE SPACE

context, spatial solution of terminal building intends to collect, translate and transform traditions of building techniques from surrounding urban context, in order to recreate the physical experience of being inside the woods.

设计为延续空间以及其外部自然环境的沟通，在空间的处理方式上意图收集，诠释，转换周边城市纹理中的传统建筑手法，从而制造身处树林的真实体验。

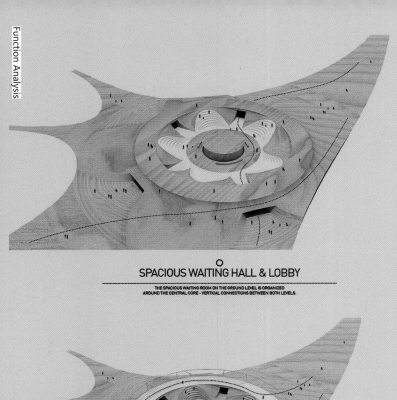

○
SPACIOUS WAITING HALL & LOBBY
THE SPACIOUS WAITING ROOM ON THE GROUND LEVEL IS ORGANIZED
AROUND THE CENTRAL CORE - VERTICAL CONNECTIONS BETWEEN BOTH LEVELS.

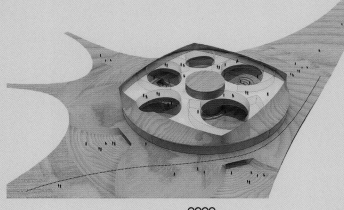

○○○○
ADMINISTRATION / CUSTOMS / OFFICES
THE DESIGN FEATURES EXTENSIVE VISUAL ACCESSIBILITY BETWEEN LEVELS, TO CREATE CONNECTIONS
BETWEEN THE PUBLIC SPACE ON GROUND LEVEL AND PUBLIC SERVICE SPACE ABOVE IT.

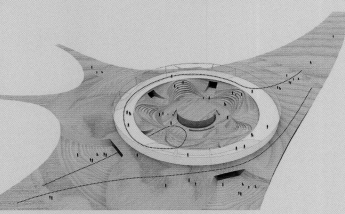

○○
COMMERCIAL / LEASE OFFICES / GALLERY
THE NEW TERMINAL PROPOSAL HAS BEEN DESIGNED AS
A TWO STORY BUILDING ACCOMMODATING LARGE SCALE PUBLIC SPACES ON THE GROUND LEVEL.

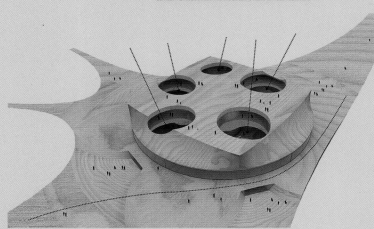

○○○○○
DAYLIGHT ACCESSIBILITY THROUGH LEVELS
THE DESIGN FEATURES NATURAL DAYLIGHT ACCESSIBILITY
BETWEEN THE PUBLIC SPACE ON GROUND LEVEL AND PUBLIC SERVICE SPACE ABOVE IT.

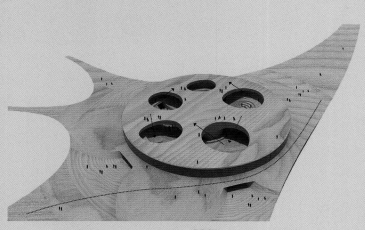

○○○
VISUAL ACCESSIBILITY BETWEEN LEVELS
THE DESIGN FEATURES EXTENSIVE VISUAL ACCESSIBILITY BETWEEN LEVELS, TO CREATE CONNECTIONS
BETWEEN THE PUBLIC SPACE ON GROUND LEVEL AND PUBLIC SERVICE SPACE ABOVE IT.

○○○○○○
TECHNICAL / MECHANICAL LEVEL
MOST OF THE TECHNICAL ROOMS ARE PLACED
ON THE TOP OF THE BUILDINGS TO PROVIDE THE BEST MAINTENANCE OPTIONS.

Design strategies
Terminal's functional layout

The new terminal proposal has been designed as a two-storey building accommodating large scale public spaces on the ground level, while the station has been designed to serve personnel and employees' offices, occupying the top level of the building. The spacious waiting room on the ground level is organised around the central core, with vertical connections between both levels. The design features extensive visual accessibility between levels, to create connections between the public space on ground level and public service space above it. In terms of function, the terminal's spatial functionality is deeply rooted into the conceptual functionality as most of the technical rooms are placed on the top of the building to provide the best maintenance options.

设计策略
终点站的功能布局

新终点站方案设计为双层建筑，其中包含了底层大范围的公共空间以及顶层私用和员工的办公空间。底层宽敞的等待室演变成整体建筑的核心，用纵向流线巧妙连接着两层建筑空间。整体的设计特点利用其独有的视觉开放性，延续了地面公共空间与上层后勤空间的连贯性。终点站每处的空间布局都围绕着概念发展，就拿功能来讲，大部分的技术后勤都放置于顶楼以达到最佳的维修需求。

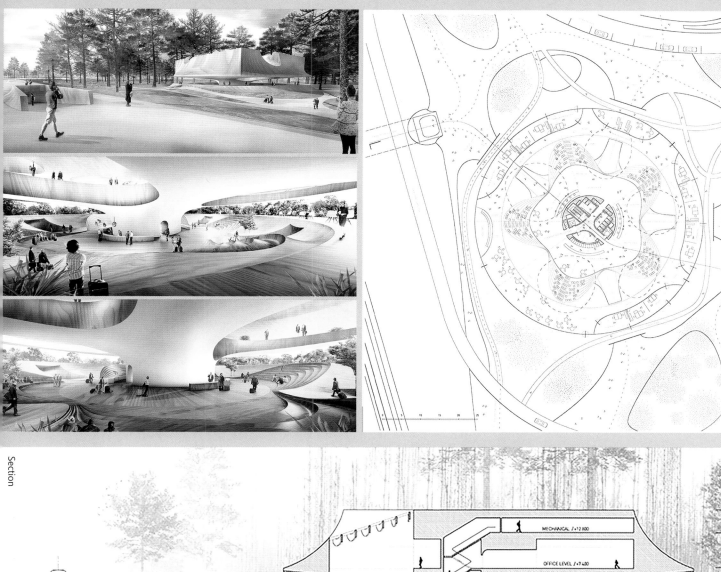

Section

West Elevation

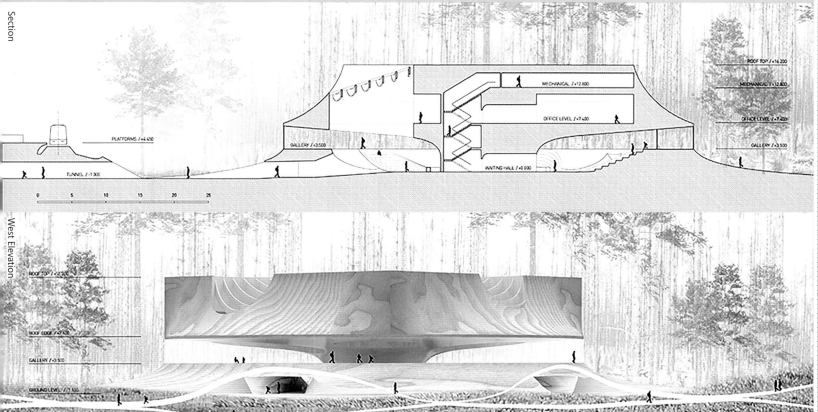

Building structure design

The new terminal proposal has been designed taking full consideration into the potential of traditional sustainable materials available in the local markets, especially wood. The importance of wood within the designed building structure is based on the historical significance of wood within the Estonian culture as well as local craftsmen's fine building techniques. Oversized laminated wood beams combined with laminated wood columns would provide a general structural anatomy for the building, while the use of light-weight steel structures will shape the rest of it. The terminal building will have a strong concrete structural base integrated into the landscape relief.

建筑结构设计

新终点站方案设计充分考虑了传统环保材料的潜质以及它在当地市场中的可及性，尤其是木质材料。木料的重要意义源于爱沙尼亚历史文化以及当地手艺人精湛的建筑技艺，因此，此次方案设计中重点使用了木质材料。大型的复合木梁和木柱将作为设计的主体支撑，轻便的钢骨则会用为辅助结构。除此之外，终点站的底座将会用水泥结构延伸到地景之中作为结构支撑。

PROBLEM / 问题

1. **RETHINK THE TOWER TYPOLOGY**
2. **RESPOND TO COMPLEX SITE**
3. **CREATE NEW TYPES OF CIVIC SPACE**
4. **INCORPORATE SUSTAINABLE DESIGN INTO TOWER FORM**

1. 重新思考塔类型
2. 应对复杂地点
3. 构造新型市民空间
4. 在塔型中纳入永续设计

OCT Tower
华侨城大厦

Studio Link-Arc
works
Studio Link-Arc
作品

Architect: Studio Link-Arc
Client: OCT Group
Location: Shenzhen, China
Area: 149,980 m²
Function: Office building

设计公司：Studio Link-Arc
客户：华侨城地产
地点：中国深圳
面积：149 980 平方米
功能：办公楼

DESIGN REQUIREMENTS
设 计 要 求

Studio Link-Arc's proposal for the OCT Tower originated from a detailed site analysis. Defined by an existing office building, a nearby tower, a sculpture park, and a major highway, the design needed to reconcile many different conditions. Rather than creating a typical urban object-tower, Studio Link-Arc's scheme seamlessly integrates architecture, landscape, and urbanism to create a contemporary public space, as well as a new tower typology that includes amenities for workers and a significantly improved interior environment.

Studio Link-Arc 的 OCT 塔提案,来源于详细的地址分析。该建筑毗邻现有办公建筑、附近塔式建筑、雕塑公园和主要公路,因此该建筑需要协调多种不同状况。Studio Link-Arc 的设计将建筑、景观和城市生活无缝衔接,以营造一种现代的公共空间以及新型塔结构,并且可为工人提供便利设施,且大大改善内部环境。

① 升高塔并延伸平台：延伸平台，以便与附近的 Hantang 广场相连，从而构建一种宽阔的广场入口，并提高升高的公共连接设施。

② 伸长塔，并增加景观：改变塔地板形状，便于光线和视野进入办公空间。在平台顶部增加景观，以便为游客和居民创建新型公共便利设施。

③ 将公园延伸到塔：在塔的南侧设置公园，从而可以将公园从地面高度延伸到塔的高度，并为办公人员提供便利设施。这些室外公园在夏季可以提供树荫，在冬天可以保温，从而可以改善用户舒适性和改善建筑的环境影响。

④ 增加防晒保护：在南面提供遮阴，从而改善能量效率，并通过减少太阳能吸收，增加用户舒适度。

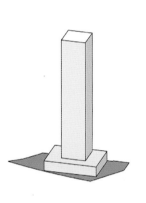

CONVENTIONAL TOWER AND PODIUM
常规塔楼和裙房

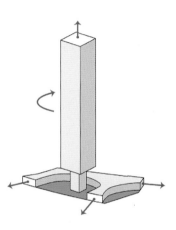

RAISE THE TOWER, EXTEND THE PODIUM
延展群楼并将塔楼提起

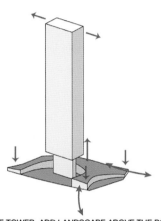

ELONGATE THE TOWER, ADD LANDSCAPE ABOVE THE PODIUM
延长塔楼 在裙房楼顶设城市景观公园

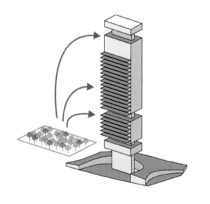

EXTEND PARK INTO THE TOWER
将绿地公园引入塔楼

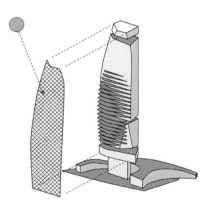

ADD SUN PROTECTION TO THE TOWER
塔楼的双层幕墙

Project Overview / 项目概况

Studio Link-Arc's proposal for the OCT Tower originated from a detailed observation of the site. Defined by an existing office building and bordered by the nearby Hantang tower, an existing sculpture park, and a major vehicular thoroughfare, the design needed to reconcile many different conditions. Early in the design process, SLA decided to avoid typical urban

我们对华侨城大厦项目设计思考源于实地对现有华侨城总部办公楼、汉唐大厦和雕塑公园、及周边环境的观察和理解，和协调这些不同方面因素的需要。在设计前期，ＳＬＡ决定避免缩小城市空间与公共景观的典型城市开发模式，而选用一种新的方法。希望通过我们的设计升华现有的景观资源，将建筑、景观与城市牢牢地结合在一起。为深圳这个城市、华侨城地区的市民、以及在此办公的员工提供一个丰富的空间节点。

Raise the Tower and Extend the Podium: Extending the podium across the site allows it to: connect to Hantang Plaza nearby, create a generous entry plaza, and create raised public connections.

Elongate the Tower and Add Landscape: Changing the shape of the tower floor plate allows the office spaces more access to light and views. Adding a landscape to the top of the podium creates a new public amenity for visitors and residents.

Extend Park into Tower: Adding gardens on the south facade allows the park to extend from the ground level into the tower, adding amenities for the office workers. These outdoor gardens create cooler shaded spaces in the summer and passively warmed zones in the winter, enhancing user comfort and enhancing the building's environmental response.

Add Sun Protection: A shaded southern facade improves energy efficiency and increases user comfort by reducing solar heat gain.

structure Analysis

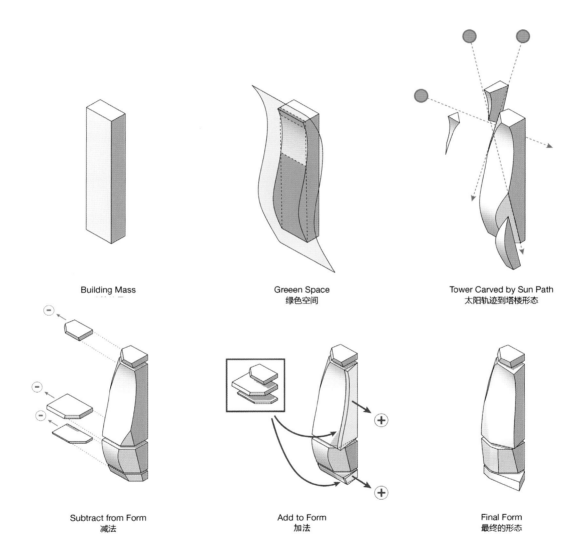

Building Mass | Greeen Space 绿色空间 | Tower Carved by Sun Path 太阳轨迹到塔楼形态

Subtract from Form 减法 | Add to Form 加法 | Final Form 最终的形态

development models which create cramped urban spaces and diminish public landscape, opting instead for a new approach. Studio Link-Arc's scheme for the OCT Tower seamlessly integrates architecture, landscape, and urbanism, creating a contemporary public space for residents, visitors, and office workers.

Elevation Analysis

幕墙系统 南向竖向遮阳
FACADE SYSTEMS - SOUTH HORIZONTAL SHADE

幕墙系统 南向横向遮阳
FACADE SYSTEMS - SOUTH VERTICAL SHADE

幕墙系统 南向整合
FACADE SYSTEMS - COMBINED

structure Analysis

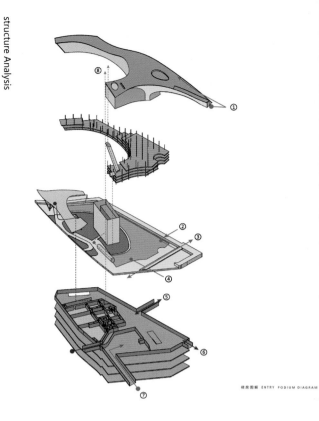

裙房图解 ENTRY PODIUM DIAGRAM

Layout Analysis

- CLUB HOUSE
- SKY LOBBY
- STANDARD OFFICE FOR CLIENT (23 FL)
- SKY LOBBY
- OFFICE FOR RENTING (15 FL)
- SKY LOBBY
- OFFICE FOR SALE (14 FL)
- LOBBY
- RETAIL
- PARKING

GREEN SPACE

Legend:
- CLUB HOUSE
- OFFICE FOR CLIENT
- OFFICE FOR SALE
- OFFICE FOR RENTING
- LOBBY
- GREEN SPACE
- RETAIL
- PARKING

功能分析 PROGRAM ANALYSIS

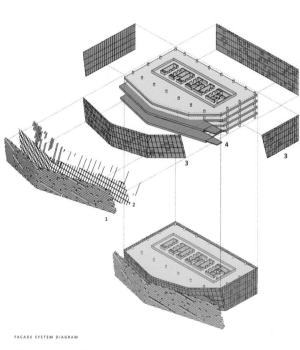

FACADE SYSTEM DIAGRAM

幕墙 FACADE NOTES: 1. 横向遮阳 HORIZONTAL SHADE LOUVERS 2. 竖向遮阳 VERTICAL SHADE LOUVERS 3. 隔热低辐 玻璃 INSULATED LOW-E GLAZING SYSTEM 4. 室外公园空间 EXTERIOR GARDEN SPACE

SLA's first design move was to lift the tower from its podium. This gesture allows daylight to reach deep into the commercial spaces, and creates more public space at grade. The roof of the podium is then articulated as a public landscape and is extended across the site in order to connect to the Hantang plaza to the east, create a generous entry plaza, and to create a

华侨城大厦的设计始于将塔楼从它的裙房中抬起，由此将允许更多的自然光线进入地面的公共空间；同时裙房屋面上形成了景观屋面，连接了汉唐广场和东面宽敞的入口广场。这个新的景观绿化屋面将提供一个新的室外庭院和雕塑公园供市民和办公楼员工共享。通过将办公楼大堂抬高在裙房之上的行为，我们创立了一个新的动线——把人从地面景观吸引上来，同时也为办公楼提供了更具可变性的室内、外空间层级。

Site Plan

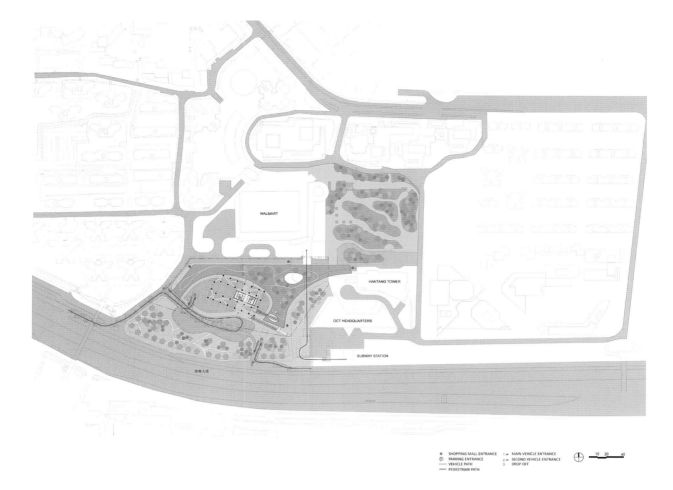

Floor Plan

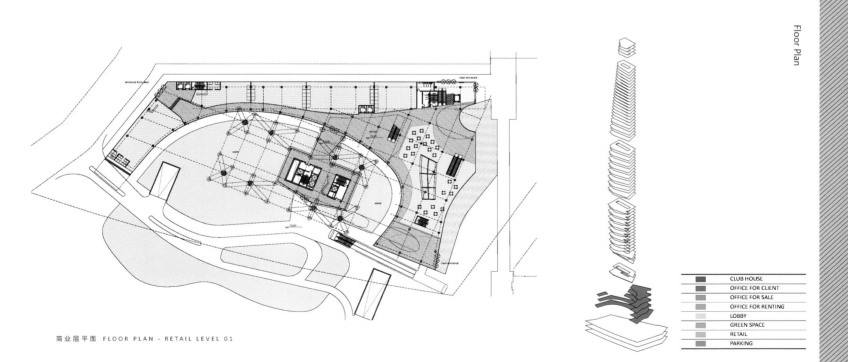

商业层平面 FLOOR PLAN - RETAIL LEVEL 01

narrower building depth that promotes natural lighting. This landscaped roof provides a new outdoor garden and sculpture space for visitors, residents, and office workers to enjoy. The office lobby is located atop this landscaped roof, enlivening the park and creating a ceremonial entrance to the tower above.

低区平面 LOW RISE PROGRAM AREA

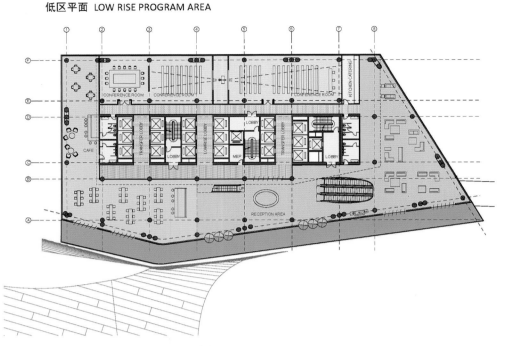
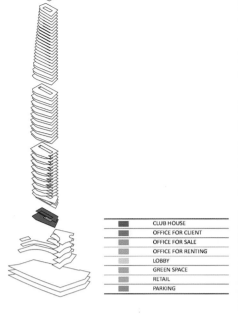

入口大堂平面 FLOOR PLAN - LEVEL 01 : ENTRY LOBBY

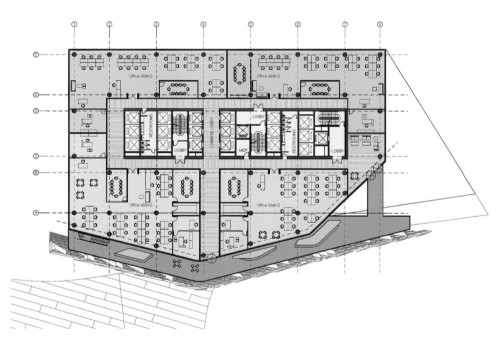
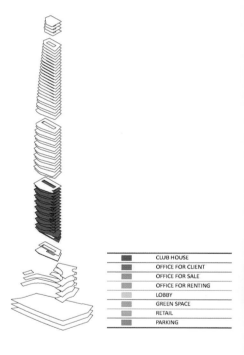

低区标准层平面 FLOOR PLAN - LEVEL 11 : TYPICAL LOW RISE OFFICE

The addition of gardens on the south facade allows the park to extend from the ground level into the tower, adding amenities for the office workers. These outdoor gardens create cooler shaded spaces in the summer and passively warmed zones in the winter, enhancing user comfort and enhancing the building's environmental response. In addition, the office tower is punctuated by a series of sky lobbies, creating extensive outdoor areas that allow workers to take breaks and conduct informal meetings.

塔楼南侧立面内部附加的庭院系统,将水平向的地面雕塑公园"生长"至竖向的办公空间内部,使员工在办公楼内部就可以直接享受到自然景观。这些户外的庭院为办公空间设立了一个自然缓冲区,为建筑在夏季提供一个凉爽的阴影空间,在冬季将阳光的温暖存留下来。
此外,空中大堂里,宽敞的室外区域为员工提供了休憩和进行非正式室外会议的机会。

中区平面 MIDRISE PROGRAM AREA

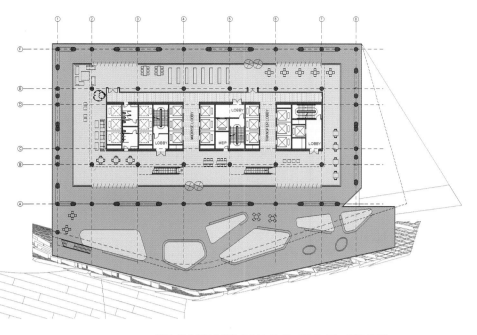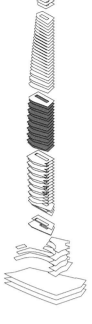

20层 空中花园层平面 FLOOR PLAN - LEVEL 20 : SKYLOBBY

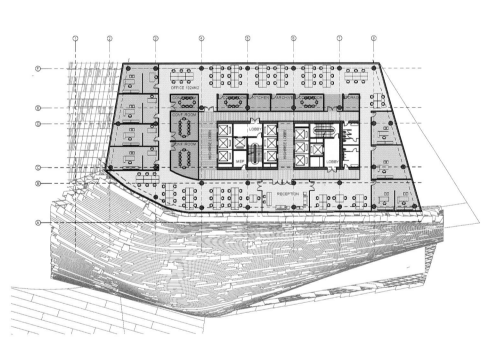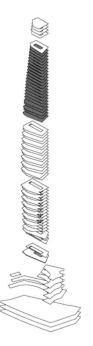

53层 高区标准层办公室平面 FLOOR PLAN - LEVEL 53 : TYPICAL HIGH RISE OFFICE

The depth of the floor plate reduces as the floors rise, allowing natural light to penetrate deeply into the office spaces while allowing for stunning panoramic views. A shaded southern facade improves energy efficiency and increases user comfort by reducing solar heat gain. The top of the building includes a clubhouse, with an interior garden that enjoys panoramic views of the city. A restaurant and lounge on the lower levels of the clubhouse is designed to maximise these views.

在办公楼的高层区域,我们减少了塔楼的进深,自然光线可以更好地渗透进办公空间,同时也为办公楼层提供了令人惊叹的城市全景。南立面的条状遮阳系统减少了太阳热量对建筑内部的辐射,提高了能源效率并使得办公室空间更加舒适宜人。在顶部的会所中,我们设计了一个新的私密庭院并拥有全景视野——建筑表皮微微折起,为庭院提供了完美的南向观赏角度。会所下层的餐厅和休息室也将提供沿窗坐席,供各个方向的景观观赏。

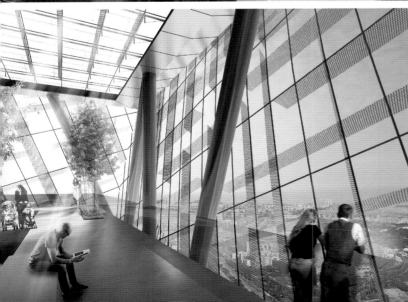

PROBLEM / 问题

1. **LANDMARK AND IDENTITY**
2. **LIMITED PREMISE**
3. **FUNCTION**
4. **HISTORICALNESS**

1. 标志性和识别性
2. 局限性的场地
3. 功能性
4. 历史性

Lavenue Crown
Lavenue crown

Atkins
works
Atkins
作品

Architect: Atkins
Location: Ho Chi Minh City, Vietnam
Gross Floor Area: 66,000 m²

设计公司：Atkins
总建筑面积：66 000 平方米
地点：越南

DESIGN REQUIREMENTS
设 计 要 求

The great vision of the clienet Lavenue Investment Corporation is to add a new landmark to Ho Chi Minh City.

客户 Lavenue Investment Corporation 的宏伟愿景，是希望为胡志明市的天际添加新的地标建筑。

THE SOLUTION 解决方式

① 建筑的塔楼以越南的国花——莲花为意象，四个花瓣状的外立面垂直向上，形成四个优美的曲面，在建筑顶部延伸成天蓬，遮蔽阳光和雨水，将会成为胡志明市天际线上一个全新地标建筑。

② 在局限的现场为客户提供一个开阔的酒店入口，宽敞的公众空间以及上落客区，同时保证楼体的设计美观优雅，并达到楼宇总面积最大化。

③ 建筑将设一家精品零售商场、一座酒店式公寓和一处华丽的空中酒吧。

④ 酒店式公寓以及屋顶酒廊为顾客提供欣赏胡志明市景色的完美地点。

Concept

1 Concept

Design Principles

Lotus flowers are beautiful in appearance. The design principle is form driven by the magnificence of the Lotus flower and in turn, stands proudly as an iconic gesture to Ho Chi Minh City. The building's facade is representative as the bud and the recess' stand as leaves. Consequently, the ancient importance of the flower is symbolised through design in present day Ben Nghe.

 →  →  →

1 Concept

Design Approach

The Lotus flower is an organic shape, free flowing and without constraints. The form of the tower is organic and free flowing, but it is simply created by a series of consistent polygonal shapes with chamfered corners.

Project Overview / 项目概况

Project overview

The tower's elegant lotus flower form is a reflection of the client, Lavenue Investment Corporation's ambitious vision for this flagship project, set to become an iconic landmark on the emerging skyline of Ho Chi Minh City. Located at one of the most important junctions on Le Duan Boulevard, close to the city's cathedral and the historic City Post Office, the 36-storey

项目概况

塔楼优雅的莲花造型折射出它的客户尚嘉投资公司对这个旗舰项目的勃勃雄心，即要使其成为胡志明市新兴天际线上的形象地标。项目座落在李杜安大道上一处最重要的交叉口，紧邻城市大教堂和老邮局。这座三十六层的建筑完工后将

The tower adopts Lotus – the national flower of Vietnam – as its image, with four petal-shaped elevations vertically upwards, forming four beautiful curved sides, which further extend into a canopy on top of the building. This canopy provides a shelter against sunlight and rain. The building will become a new skyline landmark of Ho Chi Minh City.

Provide a broad hotel entrance, special public area and drop-off area within the limited premise. Meanwhile, ensure a beautiful and elegant building design and maximise the building's area.

The building will house a retail supermarket, a hotel apartment and a top hanging bar.

The hotel apartment and bar on roof serve as perfect place for enjoying the landscape of Ho Chi Minh City.

Massing

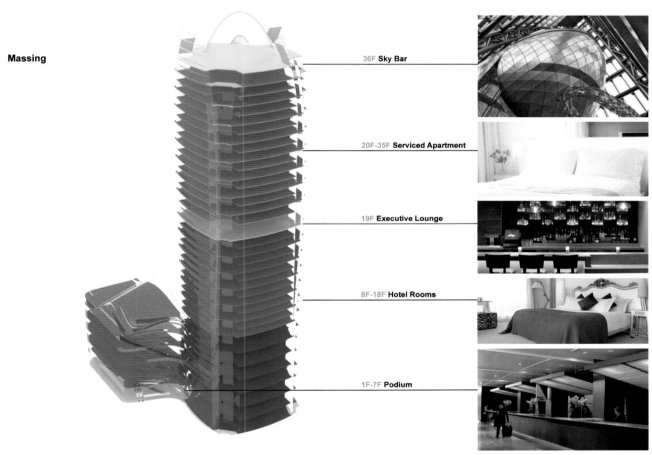

- 36F **Sky Bar**
- 20F-35F **Serviced Apartment**
- 19F **Executive Lounge**
- 8F-18F **Hotel Rooms**
- 1F-7F **Podium**

Lavenue Crown Project Concept Design Final Presentation

development will become home to a luxury five-star Langham hotel, a boutique retail mall, serviced apartments and a spectacular sky bar upon completion. The tower is approximately 160 metres high with a total gross floor area of 66,000 square metres and takes full advantage of the excellent views in almost all directions. It features four petal-like facades that extend down to provide a series of canopies and up to shelter the roof terraces of the sky bar.

用作奢华的五星级朗庭酒店的主楼，将设一家精品零售商场、一座酒店式公寓和一处华丽的空中酒吧。塔楼高约160米，总建筑面积66 000平方米，各个方向都有极好的视景。它的四个花瓣状的外立面像帽子一样向下延伸形成天篷，遮蔽了屋顶露台的空中酒吧。

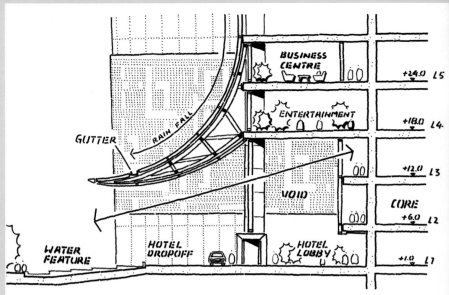
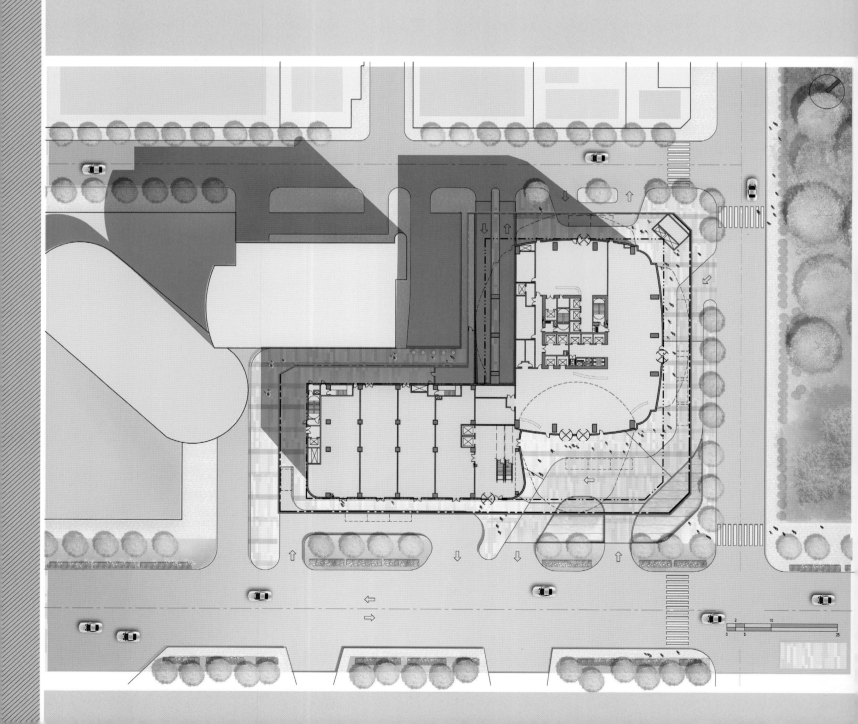

PROBLEM / 问题

1) <u>Features</u>
2) <u>ARTISTIC</u>
3) <u>ENERGY-SAVING</u>

<u>1. 功能性</u>
<u>2. 艺术性</u>
<u>3. 节能性</u>

Museum of Contemporary
Art & Planning Exhibition,
Shenzhen

深圳市当代艺术馆与城市
规划展览馆

COOP HIMMELB(L)AU
works
COOP HIMMELB(L)AU
作品

<u>Architect: COOP HIMMELB(L)AU</u>
<u>Client: Shenzhen Municipal Culture Bureau, Shenzhen,</u>
<u>Shenzhen Municipal Planning Bureau, Shenzhen, China</u>
<u>Location: Shenzhen, China</u>
<u>Area: 180,000 m²</u>
<u>Function: Exhibition</u>

<u>设计公司：蓝天组建筑事务所</u>
<u>客户：深圳市文体旅游局</u>
<u>深圳市规划和国土资源委员会</u>
<u>地点：深圳市</u>
<u>面积：80 000 平方米</u>
<u>功能：展览馆</u>

DESIGN REQUIREMENTS
设计要求

The Museum of Contemporary Art & Planning Exhibition (MOCAPE) is part of the master plan for the Futian Cultural District, the new urban CENTRE of Shenzhen.

深圳市当代艺术馆与城市规划展览馆(两馆)是深圳新市中心——福田区总体规划的重要组成部分。

① 该建筑的设计，需要与周围和现场特定条件相匹配，也需要一个完全不同的新造型。

② 我们的项目将两栋独立的建筑物结合在了一起：当代艺术博物馆（MOCA）和规划展览馆（PE）。两座博物馆都有其独特的作用和艺术要求。两者设计为单独的建筑实体，却仍然合并为一个完全统一的建筑体，包围在多功能建筑外墙中。该外墙透明，采用了复杂的内部照明，从外即可看到内部深处两个建筑相接的入口和过渡区域。

③ 建筑的技术设备的设计是为了减少对外部能源的需求：无污染系统，以及设施通过太阳能和地热（带地下水制冷系统），使用可再生能源，仅采用高能源效率的系统。博物馆屋顶将太阳光线过滤进展览区域，这样一来就减少了对人工照明的需求。

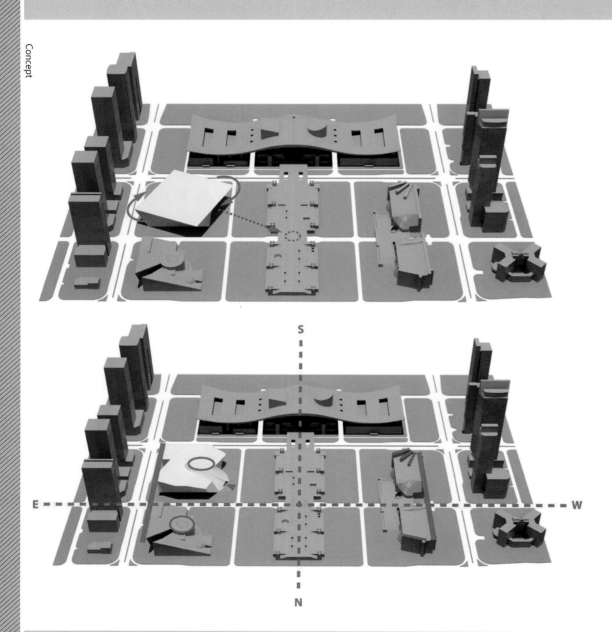

Project Overview / 项目概况

Project overview

The project combines two independent yet structurally unified institutions: The Museum of Contemporary Art (MOCA) and the Planning Exhibition (PE) as a cultural meeting point and a venue for architectural exhibitions. The lobby, multifunctional exhibition halls, auditorium, conference rooms and service areas will be used jointly.

项目概况

本项目包含了两个独立的但共享同一个建筑的机构，即当代艺术馆和城市规划展览馆，分别作为文化交汇之处以及建筑展览场所。大厅、多功能厅、礼堂、会议室和服务区域将由两者共同使用。

Design a building that matches the surrounding and site-specific conditions and yet has a completely new form.

The project combines two independent institutions: The Museum of Contemporary Art (MOCA) and the Planning Exhibition (PE). Both museums have their individual functional and artistic requirements. They are designed as separate entities but still merged in a monolithic body surrounded by a multifunctional facade. This transparent facade and a sophisticated internal lighting concept allow a deep view into the joint entrance and transitional areas between the buildings.

The technical building equipment is designed to reduce the overall need of external energy sources: Pollution-free systems and facilities use renewable energy sources through solar and geothermal energy (with a ground water cooling system) and only systems with high energy efficiency have been implemented. The roof of the museum filters daylight for the exhibition rooms, which reduces the need for artificial lighting.

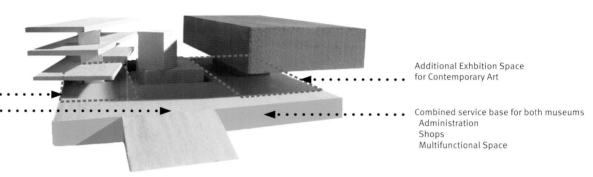

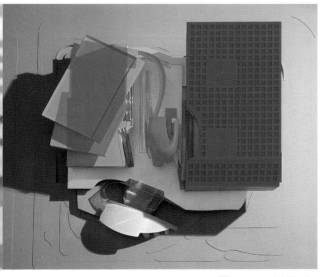

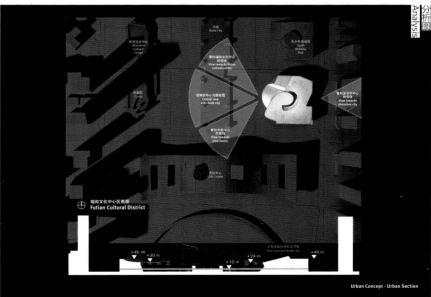

Both museums are designed as separate entities emphasising their individual functional and artistic requirements and yet are merged in a monolithic body surrounded by a multifunctional facade. This transparent facade and a sophisticated internal lighting concept allow a deep view into the joint entrance and transitional areas between the buildings. From the inside, visitors are granted an unhindered view onto the city suggesting they are somewhere in a gently shaded outdoor area, an impression enhanced by 6-to-17-mete-high, completely open and column-free exhibition areas.

艺术馆和展览馆被设计成独立的个体,强调二者独特的功能和艺术要求,然而在建筑上两者却形成了一个整体,由一个多功能的立面环绕。这个透明的立面以及先进的内部灯光设计概念使人能从外部就看到联合入口处以及两部分之间的过渡区域。同时,参观者可以从内部一览城市的景色,仿佛他们正置身于柔和阴影笼罩下的户外空间,尤其是6米至17米高的完全开放的无柱展览空间更是加强了这一印象。

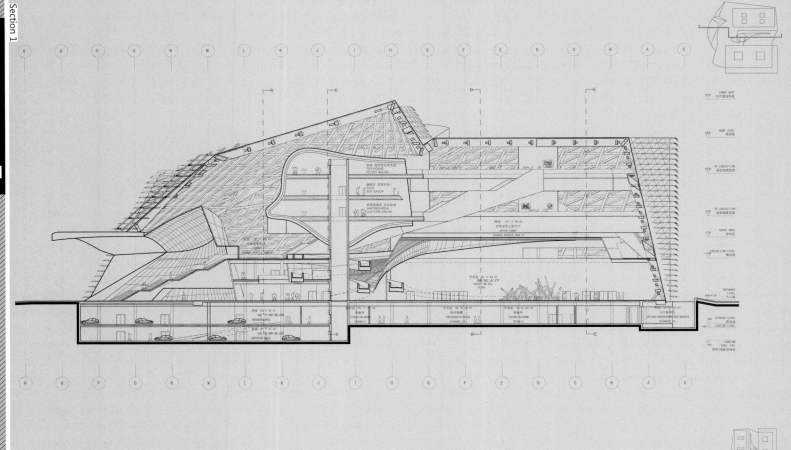

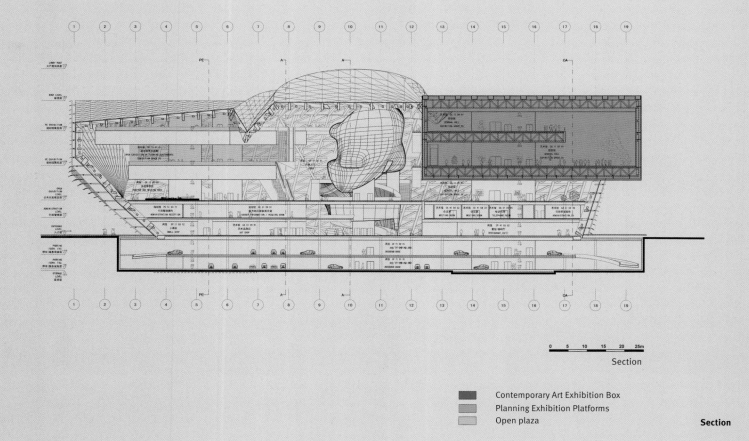

■ Contemporary Art Exhibition Box
■ Planning Exhibition Platforms
■ Open plaza

Section

Behind the entrance area between the museums, visitors ascend to the main level by ramps and escalators and enter the "Plaza", which serves as a point of departure for tours of the museums. From the Plaza the rooms for cultural events, a multi-functional hall, several auditoriums and a library can be accessed.

在两馆入口处的内部，通过斜坡和自动扶梯，参观者来到主层的广场处，这里是参观的起点。广场还通向文化活动厅、多功能厅、礼堂以及图书馆。

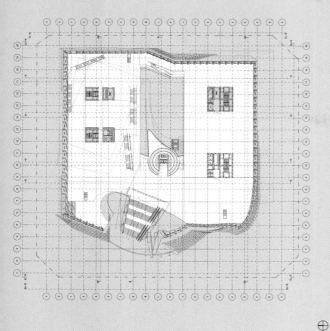

10th Floor Plan

Plan
Level +10

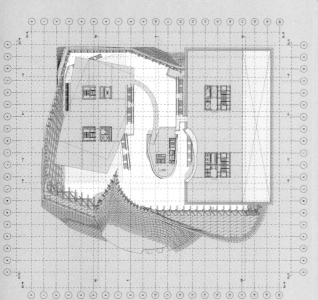

30th Floor Plan

Plan
Level +30

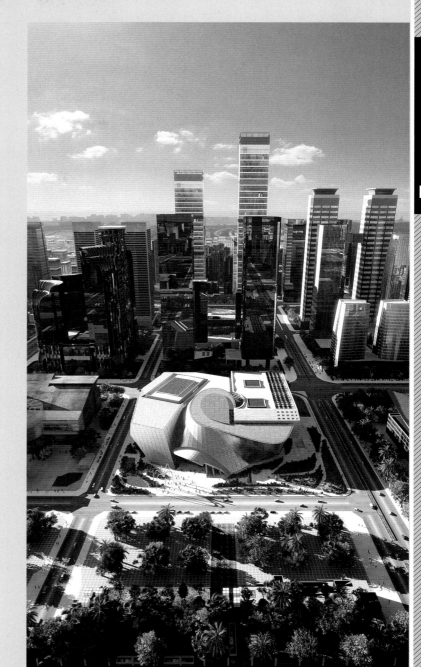

A silvery shining and softly deformed "Cloud" serves as a central orientation and access element on the Plaza. On several floors the Cloud hosts a number of public functions such as a café, a book store and a museum store and it joins the exhibitions rooms of both museums with bridges and ramps. With its curved surface the Cloud opens into the space reflecting the idea of two museums under one roof.

广场上闪着银光、形状柔和的云雕塑起到了导向和通道的作用。云雕塑内部的数层空间内容纳了咖啡厅、书店和商店，并且通过天桥和斜坡与艺术馆和展览馆分别相连。云雕塑弯曲的反光表面使它融入到空间之中，反映了艺术馆和展览馆二者同处一个屋檐下的理念。

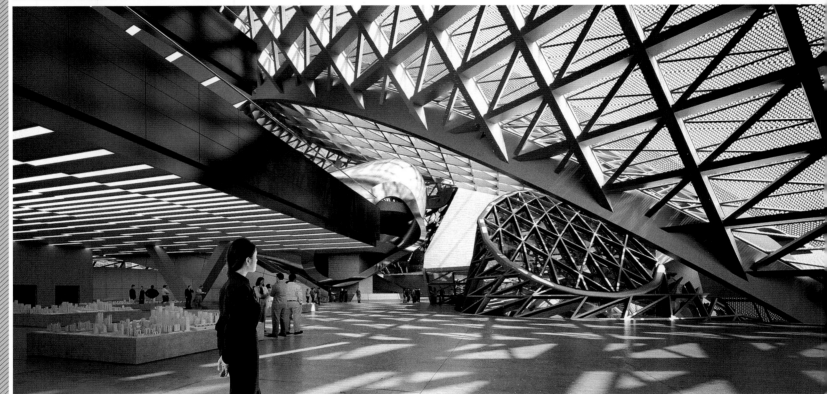

The urban concept

The MOCAPE monolith completes the eastern part of the master plan for the city centre and fills the last gap in the Futian Cultural District between the "Youth Activity Hall" (YAH) to the north and the opera-library complex to the south. Similar to other buildings in this district, the main level of the MOCAPE lies 10 meters above the ground level and so creates a stage-like platform, which acts as a unifying element with the adjacent buildings.

Skin, light and energy concept

The exterior skin consists of an outer layer of natural stone louvres and the actual climate envelope made from insulated glass. These elements form a dynamic surface, which is structurally independent from the mounting framework of the museum buildings. This functional exterior envelops the two museums, a vertical access and entertainment element (Cloud), the public Plaza, and the multifunctional base.

城市设计

两馆是对城市中心规划区域东部的一个补充,并且填补了福田文化区深圳少年宫北面以及歌剧院/图书馆南部的空白。与本区域的其他建筑类似的是,两馆的主层位于地上10米处,这创造了一个像舞台一样的视觉平台,成为其与周边建筑统一的元素。

表皮、灯光和能源设计

建筑外表皮的外层为天然石材,内层为中空玻璃,起到隔热的作用。这些元素组成了极富张力的外表皮,它与两馆建筑相对静态的空间结构相对独立。这层功能性的外表皮包裹着艺术馆与展览馆、一条垂直通道和云雕塑、公共广场以及多功能底座。

The technical building equipment is designed to reduce the overall need of external energy sources: Pollution-free systems and facilities use renewable energy sources through solar and geothermal energy (with a ground water cooling system) and only systems with high energy efficiency have been implemented. The roof of the museum filters daylight for the exhibition rooms, which reduces the need for artificial lighting.

With this combination of state-of-the-art technological components, a compact building volume, thermal insulation and efficient sun shading the MOCAPE is not only an architectural landmark but also an ecological and environmentally friendly benchmark project.

建筑机械设备的选用旨在减少建筑的整体能耗。为了达到这个目标，配置了一系列无排污的太阳能、地源能（包括地源能量制冷）可再生能源设备，而且只有那些能源效率高的设备才被采用。博物馆采用过滤日光进行照明，减少了人工光源的使用。

最先进的技术设备、紧凑的建筑体量、高效的隔热保温措施以及遮阳手段——两馆项目不仅仅将成为深圳的建筑地标，而且也将是生态和环保方面的标杆项目。

PROBLEM / 问题

① LOCATED ON A PENINSULA IN THE CONFLUENCE OF THE RHÔNE AND SAÔNE RIVERS

② AS A MEETING POINT OF NATURAL LANDSCAPE AND SCIENCE AND ART SPACE

③ MAINTAINING ENERGY BALANCE OF THE BUILDING

1. 坐落于罗纳河与索恩河交汇处的一个半岛上
2. 为自然景观与科学与艺术空间的交会点
3. 保持建筑的能源平衡

Confluence Museum in Lyon, France
法国里昂汇流博物馆

COOP HIMMELB(L)AU
works
蓝天组建筑事务所
作品

Architect: COOP HIMMELB(L)AU
Location: Shenzhen, China
Area: 195,206 m²

设计公司：蓝天组建筑事务所
地点：法国里昂
面积：195 206 平方米

DESIGN REQUIREMENTS
设 计 要 求

Right from the 2001 international competition for a natural history museum in Lyon, the museum was envisioned as a "medium for the transfer of knowledge" and not as a showroom for products.

从 2001 年在里昂举办的国际自然历史博物馆比赛开始，对博物馆的构想即为"传递知识的媒介"，而不仅仅是一间产品展示厅。

THE SOLUTION
解决方式

① 从一开始就很明显，这个地块位置很棘手。因为地下水位高，博物馆不得不拔高悬于水面之上—这就得将536跟桩牢牢地打入地下30米。

② 地点的复杂性带来了博物馆新的几何造型。我们的设计构想是，建造一座能消除与大自然之间屏障的博物馆，创造与大自然和谐共处的建筑造型的通道。

③ 在能耗方面，汇流博物馆是一座高效能建筑，仅仅22%-23%的制冷制热功能会与自然元素相抵消，如能源损失、能源输入和阳光效果

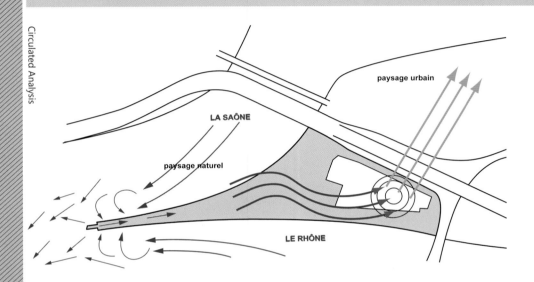

Diagram of the urban circulation

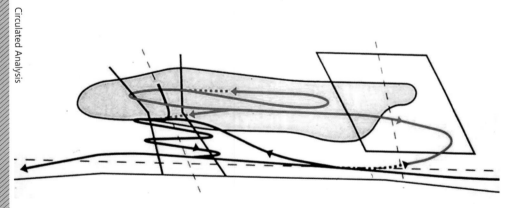

Circulation diagram

Project Overview / 项目概况

Location analysis

The building ground of the museum is located on a peninsula that was artificially extended 100 years ago and situated in the confluence of the Rhône and Saône rivers. Even though it was apparent that this site would be a difficult one (536 piles had to be securely driven 30 metres into the ground), it was clear that this location would be very important for the urban design. The building should serve as a distinctive beacon and entrance for the visitors approaching from the South, as well as a starting point for urban development.

It was apparent from the beginning that the site would be a difficult one. Due to the high water table the museum had to be lifted above grade - 536 piles had to be securely driven 30 metres into the ground.

New geometries were a response to the complexity of the location. The design envisions a museum that eliminates barriers to the natural world, creating a passageway of forms built to harmonise with nature.

The Musée des Confluences is a high performance building in terms of energy consumption as only 22-23% of its heating and cooling production is used to counteract the natural elements such as energy losses, energy inputs and effect of sunshine.

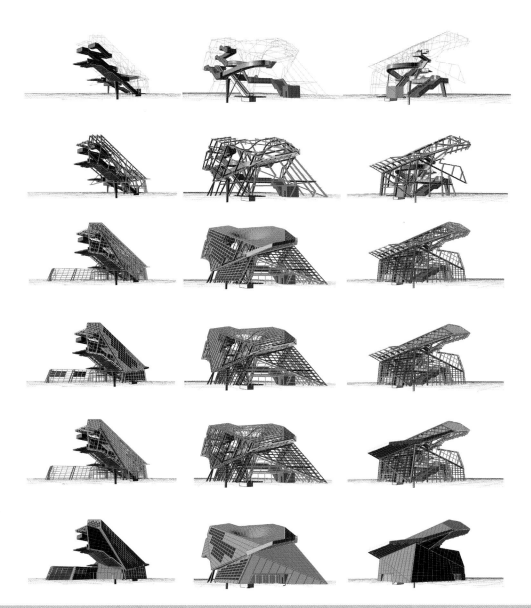

该博物馆建筑地面坐落于一个100年前人工延展的半岛上，位于罗纳河与索恩河的交汇处。尽管，很明显，该建筑场地会困难重重（536根桩基柱必须牢牢打入地面30米以下），但该位置对城市设计也将十分重要，这一点毋庸置疑。该建筑将作为一个独特的灯塔，也是从南部前来的参观者的参观入口，同时也是城市发展的起点。

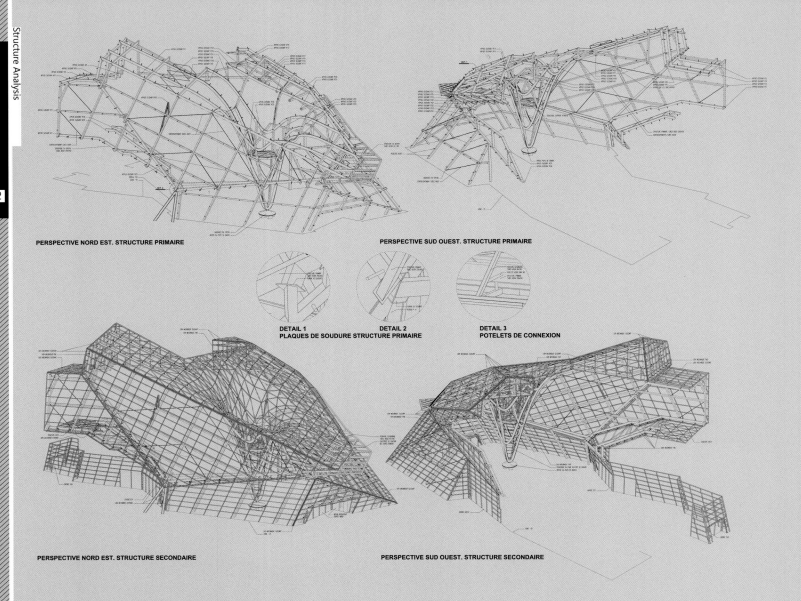

PERSPECTIVE NORD EST. STRUCTURE PRIMAIRE PERSPECTIVE SUD OUEST. STRUCTURE PRIMAIRE

DETAIL 1
PLAQUES DE SOUDURE STRUCTURE PRIMAIRE
DETAIL 2
DETAIL 3
POTELETS DE CONNEXION

PERSPECTIVE NORD EST. STRUCTURE SECONDAIRE PERSPECTIVE SUD OUEST. STRUCTURE SECONDAIRE

The Cryst

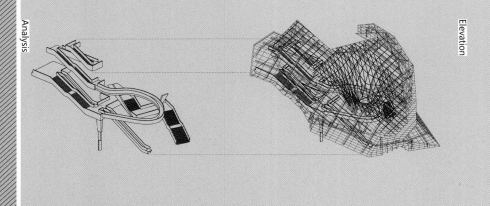

The Crystal: Circulation

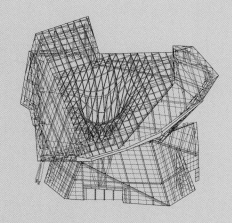

Crystal: north elevation

Design concept

This project is mainly focused on its sustainability, environmental protection idea, business and culture. However, sustainability will be an important factor to judge whether this project is successfully developed so as to create a nice living environment by showing respect towards nature. Meanwhile, innovations will be made to the community's architectural form; SGBC community is aiming at providing its renters with a place for cultural exchange and activities, and it will attract wide attention from the world.

设计理念

本案主要以可持续性、绿色环保、商务性和文化性为理念。本案在设计中将可持续发展作为衡量开发是否成功的重要标志，创造一个良好的居住环境以表达对自然的尊重；同时注重社区建筑形态的创新；SGBC 社区希望为租赁者提供文化交流和活动的场所，并且在世界范围内备受瞩目。

设计策略

1. 可持续性：
新城的开发有责任去创造一个良好的居住环境以表达对自然的尊重。SGBC 社区希望成为可持续设计的典范。从总体规划设计来看，建筑的形态主要是考虑用以引导盛行风更多的经过本基地，经过建筑的悬臂式露台以及立面上的通风系统，我们需要考虑每一个设计的细节以保证其可持续水平。新的开发将通过夏季的被动加热、保温，冬季地热自然冷却能源来确保建筑的低能耗。院落几何形态使我们增加了自然采光，以节省人工照明。由大量植被覆盖的屋顶花园将帮助收

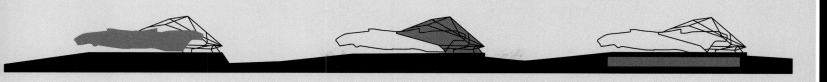

Nuage Cloud   **Cristal** Crystal   **Socle** Plinth

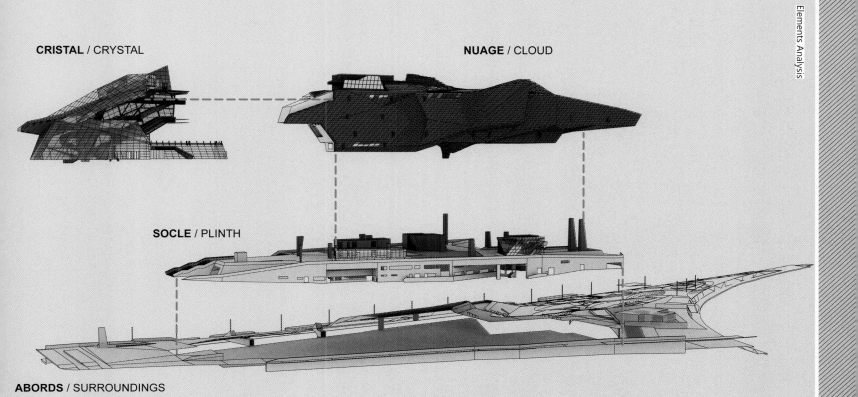

CRISTAL / CRYSTAL

NUAGE / CLOUD

SOCLE / PLINTH

ABORDS / SURROUNDINGS

The elements

Design strategies
1. SUSTAINABILITY:
The new city developments have the responsibility to create a better living environment more respectful with nature. SGBC® community wants to be a leading example in sustainable design. From a master plan design where the blocks are shaped to enhance the prevailing breezes to flow through the site, to an architectural facade design with cantilevered terraces and crossed ventilation units, the architects need to think of every level of sustainable design details. The new development will save energy through natural cooling in summer and passive heating, insulation and geothermal in winter. The courtyard typology enhances natural lighting allowing the project to save on artificial lighting. The green roofs and parks will help collect rain water that could be reused throughout the site. Pedestrian circulation through the site as well as the easy communication to the city public transportation system will minimise the usage of private cars. Photovoltaic panels can be used in facades, roofs and pergolas. In the north side of the south the architects propose a water purification pond and a wind generator park.

集雨水以促进其再利用。人行流线以网状遍布于整个基地，同时加强了基地内部的相互交流以及其于城市公共交通系统的联系，而且有助于减少私家车的使用。光伏板还可用于建筑外立面，屋顶和花园。在基地的东北一侧，我们设置了一个净水池和一个风力发电公园，用于整个基地的污水净化和能源的供给。

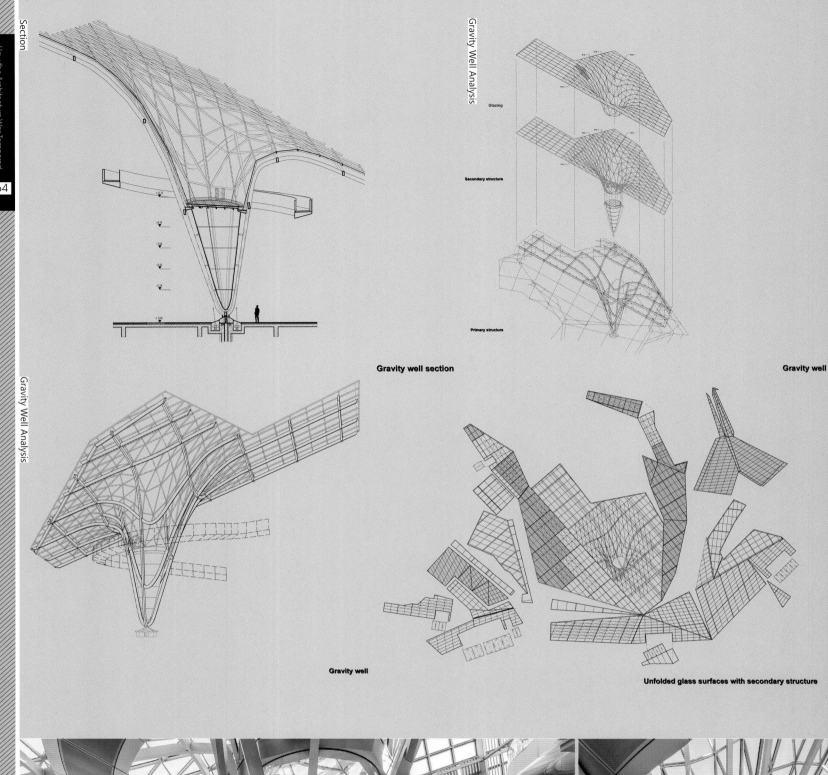

Gravity well section

Gravity well

Gravity Well Analysis

Glazing

Secondary structure

Primary structure

Gravity well

Unfolded glass surfaces with secondary structure

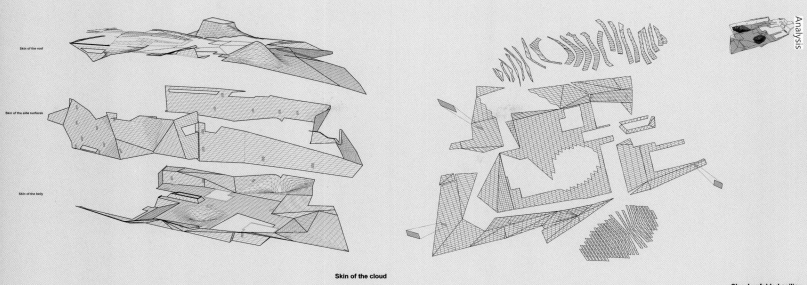

Skin of the cloud

Cloud unfolded ceiling

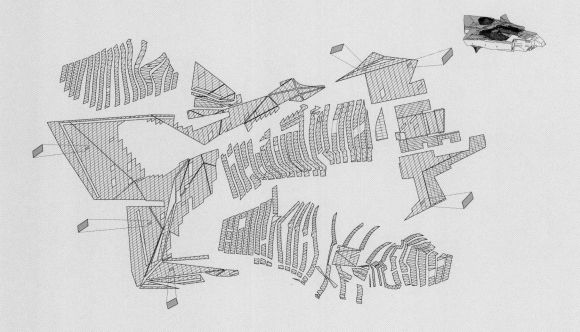

Cloud unfolded roof

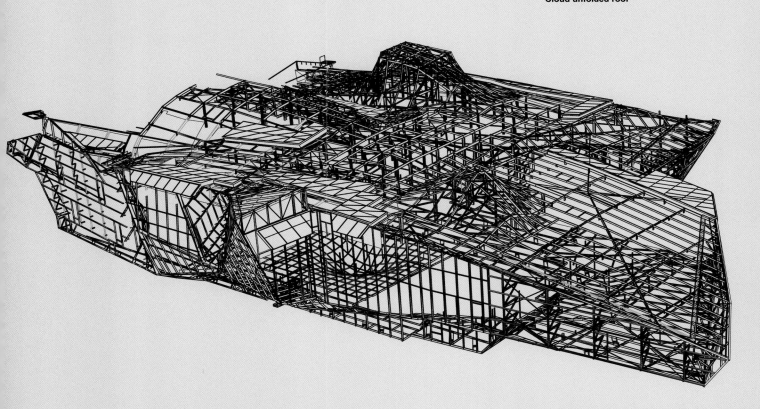

Cloud: Primary structure of the cloud

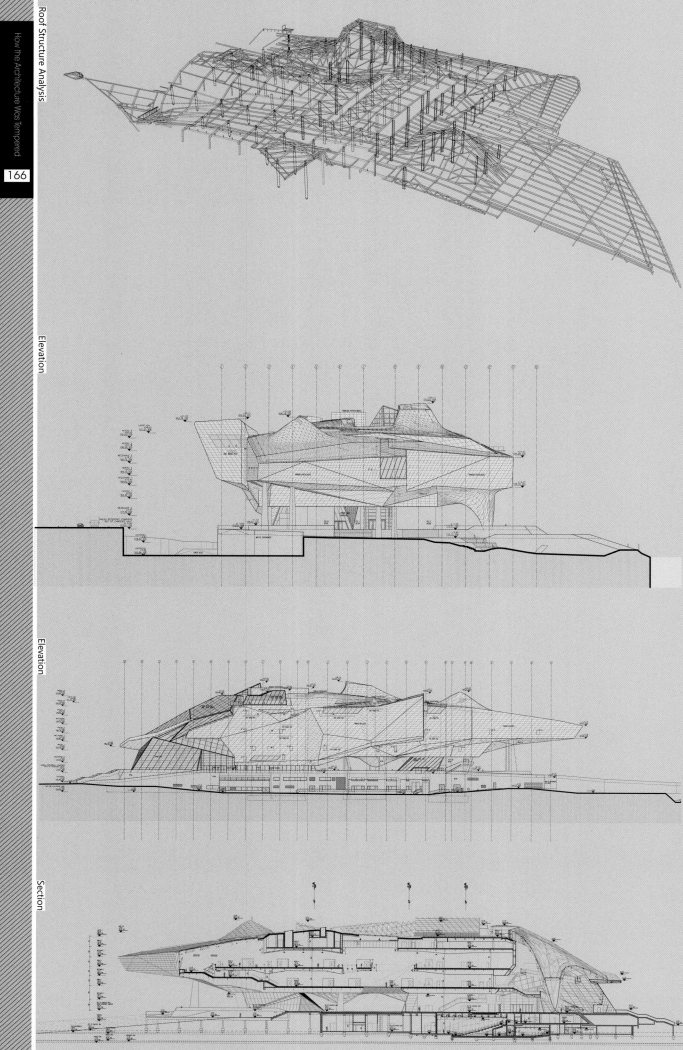

Coupe générale B-B

2. GREEN

SGBC® community is a park in itself. A 55,000m² park works as a green belt that protects the new buildings from the noise of the highway and train while hosting new equipments and open spaces for the citizens. The typology of the courtyard building allows the architects to create a big garden inside every building block allowing cross-ventilation and a double sided view from every unit. The roof tops also become green gardens. The gardens will help clean air and water as well as providing a new natural way of living inside the city.

3. BUSINESS

The new building typology of this new community is the axis of the business success. The blocks are designed to enhance the connection within the district; the rounded triangular network allows enhancing the exposure of every single building, creating perspectives that will give an extra value to every block. The courtyard building is easily dividable; the double sided facade to the street and courtyard allows for a much wider range of office and residential units, giving the possibility to always avoid the north orientation and provide natural light and sun to every unit. The Ground Floors become street retail enhancing the pedestrian communication within the site, always sheltered in the shadow of trees and cooled with the presence of fountains. The architects want to create an active community, a dynamic city where activity, business and retail can coexist. The urban and architectural designs are here focused into creating a new city typology.

4. CULTURE

As a main destination, SGBC Community will need to provide not only culture and activities to its own users, but also has to become an attraction for outside users. The history of the site with its harbour and logistic past, allows the architects to propose building a landmark tower in the centre of the site and allocate in the top floors a museum. The theme of the museum could be flexibly adapted to different uses, being a Huangpu River Museum in this proposal. The view to the river from the top of the tower and an interactive exhibition could bring the visitor to an exciting new experience.

2. 绿色

SGBC社区本身就是一个公园。一个55 000平方米的绿化带确保了整个园区可以远离从公路和铁路传来的噪音，同时提供新型的城市公共空间供广大市民休憩。院落的几何形态更可以在每栋建筑内部提供大片的相对私密的共享花园，有利于空气的对流和每个房间对景观的视觉享受。同时，在建筑的屋顶也将设置大片的屋顶花园。花园将有助于净化空气和水质，为在城市中生活的人们提供一个返璞归真的生活环境。

3. 商务

一种新型社区的建筑形态是开发是否成功的标志。基地内的每个区块的设计是为了增加每个区块之间的相互联系，圆形转角的三角形网络可以大幅度增加每个单体建筑的外露面，也就是平常所说的商业面，在消费者有限的视野内提供更多的建筑透视角度，大大提高了建筑的商业附加值。有庭院的建筑可以很容易的根据使用要求被分割成若干小的建筑单元，双侧的平面布置方式为办公和居住的人们提供了范围更广的选择，并且提供了杜绝北向的可能性，并为每个单元提供了充足的自然光线。建筑的一层将成为商业零售店，提高基地内部的商业价值和各个区块间的相互联系，并且行人可以享受到充足的树阴和广场上喷泉提供的凉爽。我们要创建一个活跃的社区，一个充满活力的城市，商务和零售可以共存。整个城市设计和建筑设计的重点就聚焦于如何创造一个新的城市类型。

4. 文化

SGBC社区的主要目的不仅是要为租赁者提供文化交流和活动的场所，更是要吸引世界各地的目光交汇于此。历史上，本项目基地过去是港口和物流站，所以我们建议建造一栋可以作为当地地标的塔楼，并在其顶层设置博物馆。博物馆的主题可以是灵活地、适应不同的用途，我们的建议是建立以黄浦江为主题的博物馆。塔楼顶部的视角和交互式的展览将吸引众多市民进行参观和互动，这将会是一次让人耳目一新的视觉盛宴。

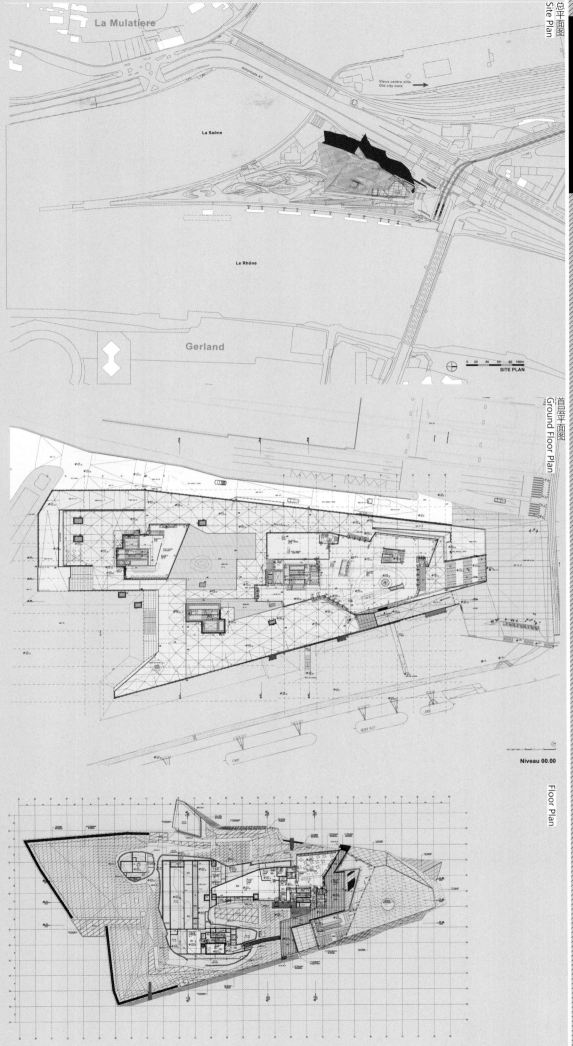

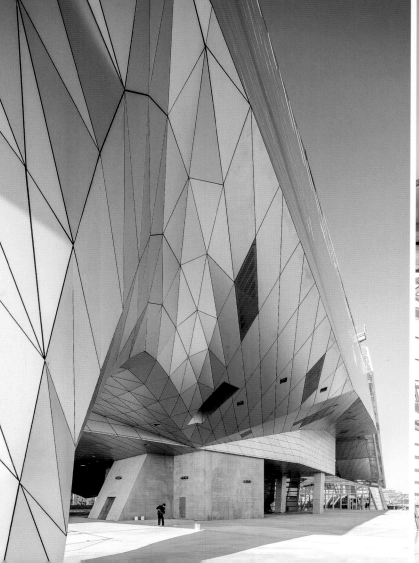

PROBLEM / 问题

1. **RESPECT FOR REGIONAL CULTURE AND HISTORY**
2. **CORRESPONDING WITH THE SURROUNDING ENVIRONMENT**
3. **ENERGY SUPPLY**

1. 尊重地域文化和历史
2. 呼应周边环境
3. 能源供应

Helsinki Central Library
赫尔辛基中央图书馆

Urban Office Architecture
works
Urban Office Architecture
作品

Architect: Urban Office Architecture
Location: Helsinki, Finland
Area: 16,000 m²
Function: Library

设计公司：Urban Office Architecture
地点：芬兰赫尔辛基
面积：16 000 平方米
功能：图书馆

DESIGN REQUIREMENTS
设 计 要 求

The project is designed to connect Helsinki City and as the library to be the natural innovation engine.

将项目设计为连接着赫尔辛基城市和用作自然创新引擎的图书馆。

① 整个建筑采用黑宝石的外形，完美的溶入到该地区的规划草案中，呼应周边现有建筑、基础设施以及城市构造。

② 通过抬高的U型建筑和一条与地面相连的坡道及屋顶露台作为周围自然公园的延伸。

③ 建筑外壳作为收集和分散热能、空气和电能的神经网络，通过一个大的传感器网络进行协调。

Sketch

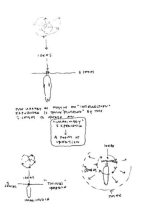
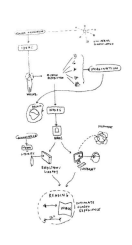
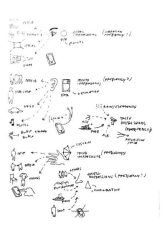
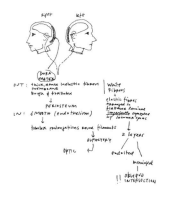

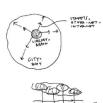
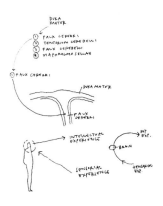
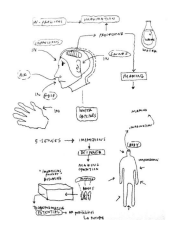
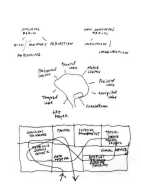

Project Overview / 项目概况

Design concept

These functions are all "kept together" by the exterior building envelope which creates a sort of continuity and consistency throughout the building. Similar to the Dura Mater of the brain, the Library envelope carries information both to its users and as a response to the environment around it.

设计理念

新赫尔辛基中央图书馆的设计概念来源于人类大脑，从喻意上来说，连接着赫尔辛基城市和可用作自然创新引擎的图书馆。正如人类大脑一样，许多区域管理着不同的功能，图书馆由不同的功能区域组成，有些区域具有完全不同的特征。

In the shape of a black stone, the whole building is perfectly blending in the draft plan of this area, corresponding with the surrounding existing buildings, infrastructure and urban structure.	**1**
An elevated buiding in the shape of U, a slope connecting the ground level and the roof terrace are the extension of the surrounding natural park.	**2**
As the neural network that collects and distributes thermal energy, air and electric energy, the building shell is coordinated through a big sensor network.	**3**

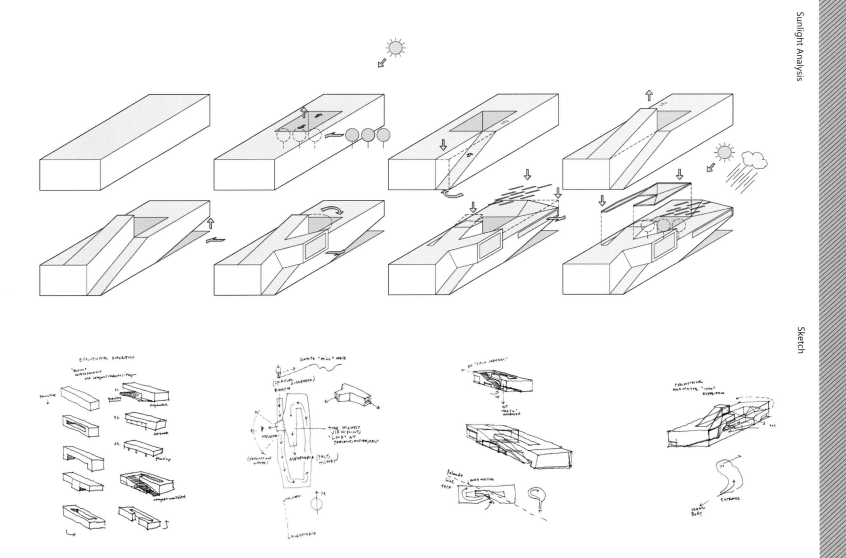

Sunlight Analysis

Sketch

The concept for the New Central Library of Helsinki is based on the Human Brain, metaphorically connecting the city of Helsinki with its natural creative engine, the Library. Like the Human Brain, whose various areas manage different functions, the library is organised in very different zones and functions, some completely opposite in character and nature.

这些功能区由外部的建筑外壳连接在一起，实现了建筑的连续性和统一性。正如大脑的硬脑膜一样，图书馆将信息传送给用户，并与周围的环境交相辉映。

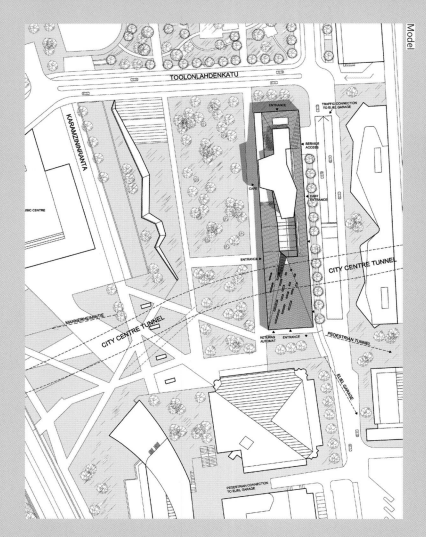

Site Plan

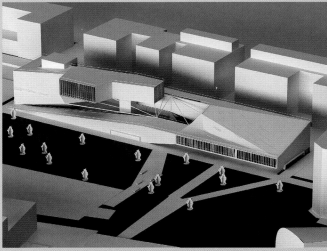

Model

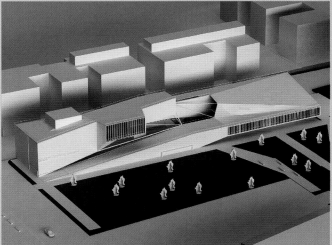

Design strategies
1. MAIN PRINCIPLES BEHIND THE ARCHITECTURE AND CITYSCAPE SOLUTION

The new Helsinki central library serves as an extension of the natural park surrounding it. The park filters itself into the library travelling from the West to the East of the city block becoming "trapped" into building structure. The park pushes the masses both horizontally, North and East, and vertical, ground to sky. The result is a U-shape building elevated thought a ramp travelling from the ground level and culminating in a roof terrace. Furthermore, the horizontal force of the park divides the building into two wings, North and East and transforms itself into an interior reading garden. Horizontally, the building is connected by the café at the ground level and the restaurant and sauna at the top level linked by a direct lift. The trapped park forms two panoramic windows seeking to escape via looking in the direction of the Parliament building to the West and Alexandria to the South. All the knowledge brought into the library by the city through the park makes the new Helsinki central library the brain of the metropolis; a direct receiver and transmitter of information clearly reflected by its constant ever-changing programme of events and its permanent collection of knowledge.

The Library acts as a receiver and emitter of information creating an "8" figure loop with the city of Helsinki and the World at large. Different from an enclosed enclave which protects knowledge, the Dura Mater Central Library collects and spreads knowledge at the broadest possible scale.

The Library is accessible on its roof via a ramp that begins at grade on the main piazza. The promenade ascension is both a physical experience of the changing cityscape and the spiritual counterpart to the idea of "change" in mind. Similar to Dante's journey, the sloping uphill can be experienced as a transformative moment, arriving at the top floor room which faces south towards Alexandria and the Finnish Parliament.

The Library ultimate view points are the two large window-rooms of the top floor restaurant, Jazz bar and Sauna. These spaces face west and south, aligning respectively with the Parliament and House of Music, and the trajectory of the Alexandria Library in Egypt. These views activate a connection to the past, present and future of the library and welcome visitors to re-experience it in a forward-thinking dimension.

2. OVERALL TECHNICAL STRATEGY: DURA MATER PERFORMANCE ENVELOPE

This building is designed to use the building envelope as the primary thermal and electrical generator and regulator and uses the district heating, cooling and grid electricity as efficient back-up. The technical systems of this building are primarily housed within the multi-layered assembly of the building envelope carrying an interior and exterior narrow planar cavity or plenum for the flow of energy and material (hydronic heat, ventilative air, DC electrical). A translucent isolative core of aerogel, luminous concrete or motorised shading is bounded on both interior and exterior by these cavities and the surface is

设计策略
1. 主要建筑原则和城市景观解决方案

新的赫尔辛基中央图书馆是周围自然公园的延伸。公园本身对这个城市从西到东前往图书馆的路线进行了规划，并受困于建筑结构。公园对建筑体量从横向的东西方向和垂直的地面到天空方向都进行了推进，因此形成了一个抬高的U型建筑，有一条坡道与地面相连并且有一个屋顶露台。另外，公园的横向力将建筑分成北翼和东翼，并将其本身转换成一个内部阅读花园。从水平方向来说，建筑低层的咖啡馆和顶层的餐厅和桑拿中心由一个直电梯进行连接。被困住的公园形成了两个全景窗户，以避免从西侧国会大厦和南侧亚历山大的直接视线。通过公园从城市带入图书馆的所有知识让新赫尔辛基中央图书馆成为这个城市的大脑，其不断变化的活动项目和永恒的知识收集能力使其成为信息的直接接收中心和转换中心。

图书馆作为信息的接收器和发射器，为赫尔辛基这个城市乃至全世界打造了一个"8"字型循环系统。不同于保护知识的封闭式飞地，这个硬脑膜式的中央图书馆可在最大的范围内收集和传播知识。

人们可以从主广场的斜坡上到屋顶，然后由此进入图书馆。上升的长廊不仅是对不断变化的城市景观的实质反映，还是心中对"变化"这个概念的精神对应。

图书馆的终极视点是顶层餐厅、爵士酒吧和桑拿中心的两间大观景室。这些空间面向西和南面，与国会大厦、音乐之家分别排成一行，并与埃及的亚历山大图书馆处于同一平面。这些景象激活了图书馆的过去、现在和未来的联系，欢迎游客从前瞻性思维的角度来重新体验这个空间。

2. 整体技术战略：硬脑膜性能结构

设计中将建筑外壳用作主要发热机和发电机以及调整器，并使用区域供暖、制冷和并网电作为有效的备用设施。这栋建筑的技术系统被设置在建筑外壳的多层架构中，建筑外附设有狭窄的内部和外部平面腔道或布满能源和材料线路（循环热能、空气通风、直流力）。透明的隔离式凝胶核心、发光的混凝土或电动遮阴系统通过这些腔道布满内部和外架构，建筑表面还覆盖着双窗格绝缘发光物质。这样复杂的架构将热能、空气和电力以细管式结构分散至建筑的所有区域并在建筑表面集成的传感器的协助下，以不断流动的自我调节方式来分散、平衡和交换建筑热能的生成和消耗，空气的排放以及电力的生成和消耗。

智能的建筑外壳在北端、南端和中部插入四个垂直的轴式架构，计划以垂直的方式实现完成，并连接至建筑中心的机械布置空间，以实现中央集中控制、热能循环、电力控制、循环热泵、通风和区域供暖和制冷的连接。

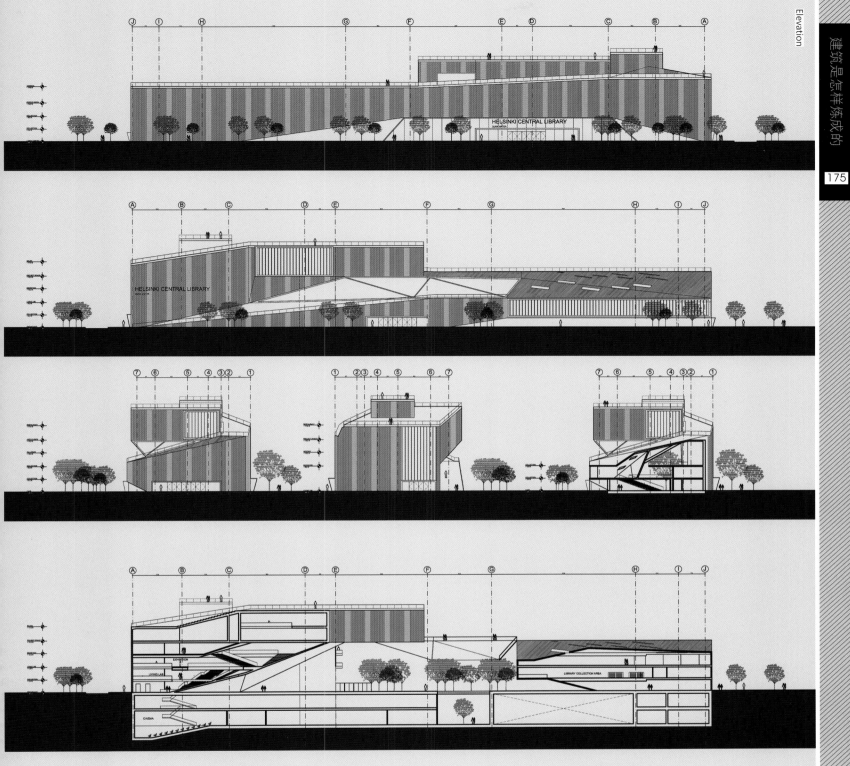
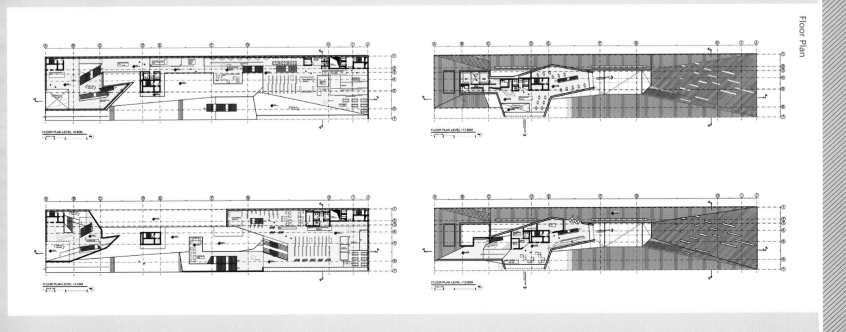

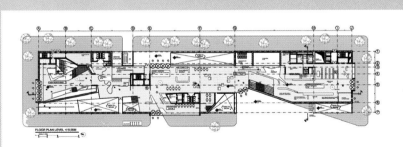
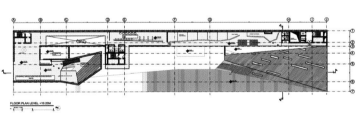

Floor Plan

enclosed with and interior and exterior dual pane insulated glazing units. This sophisticated jacket allows overall capillary-scale distribution of heat, air and electricity to all areas of the building and assists the building to distribute, balance and exchange the heat produced and heat consumed; fresh air delivered and stale air exhausted, electrical energy generated and electrical energy consumed in a constant flow of self-regulation, assisted by a full network of skin-integrated sensors.

The intelligent envelope plugs into the four vertical shafts at the north end, south end and centre of the building plan to gather distribution vertically and connect to the building centre mechanical area in order to plug into a central brain unit, heat exchange, electrical control, water-to-water heat pump, air supply and exhaust, as well as the connection to the district heating and cooling.

3.ENERGY SUPPLY SOLUTION

The building will utilise district heating and cooling along with grid electricity only when the collection, exchange and balancing capacity of the intelligent envelope requires additional energetic input. Water-to-water heat pumps will be employed to provide cooling and/or heating required for the building as the inputs required to balance the whole of the integrated system flows.

The building envelope incorporates building-integrated photovoltaic panels (BIVP) in the exterior insulated glazing unit on surfaces facing SE, S and SW to produce electrical power.

The building envelope operates as a neural network of production collection and distribution for heat, air and electricity, coordinated by a large network of sensors throughout the same surfaces. These are connected together through a central responsive logic and uses the district heating and cooling connection and grid electricity as back-up when required or more efficient than the building scale.

3. 能源供应解决方案

建筑只有在智能建筑外壳收集、交换和平衡功能需要额外的能源输入时才会使用区域供暖和制冷以及并网供电。我们将使用水循环热泵来满足建筑的制冷和／或功能需求，而且这样输入的能源可用来平衡整个集成系统的循环。

建筑外壳面向东南、南部和西南的表面上覆盖着建筑集成光伏板和外部绝缘发光材料，可为建筑生成电能。

建筑外壳作为收集和分散热能、空气和电能的神经网络，通过一个大的传感器网络进行协调。这些功能都通过中央反应式逻辑系统连接起来，并在建筑需要更有效的大规模操作时使用区域供暖、制冷和并网供电作为备用设施。

PROBLEM / 问题

1. **CREATIVE TEACHING SPACE**
2. **COMMUNICATION SPACE**
3. **MULTI-FUNCTION**
4. **OPENNESS**

Learning Centre Polytechnique
巴黎 - 萨克雷大学内巴黎综合理工学校新学习中心

SOU FUJIMOTO ARCHITECTS
works
SOU FUJIMOTO ARCHITECTS
作品

Architect: SOU FUJIMOTO ARCHITECTS
Architects Team: SOU FUJIMOTO ARCHITECTS + MANAL RACHDI OXO ARCHITECTS + NICOLAS LAISNE ASSOCIES
Location: Paris-Saclay University, France
Site Area: 10,000 m²
Function: Learning centre

设计公司：藤本壮介建筑设计事务所
设计师团队：SOU FUJIMOTO ARCHITECTS + MANAL RACHDI OXO ARCHITECTS+ NICOLAS LAISNE ASSOCIES
地点：法国巴黎 - 萨克雷大学
占地面积：10 000 平方米
功能：学习中心

DESIGN REQUIREMENTS
设计要求

The project is a new learning centre of école Polytechnique in Université Paris-Saclay. In this centre, six educational and research institutions may share their learning projects. It will also stand as an important landmark there.

项目为在巴黎-萨克雷大学内的巴黎综合理工学校新的学习中心，六所教育和研究机构可以在此共享学习项目，并作为其重要的标志性建筑存在。

THE SOLUTION
解决方式

① "远程学习房间"、以及诸如"盒子工作"或"项目房"等的视频会议和协作工作空间将最大化优化以互动和数字化工具为重点教育体系的环境。

② 创造了许多非正式的集会或工作空间,人们不会再在走廊里擦身而过,而是在生动的场所,在沐浴着柔和光线的、有着令人惊讶和不断变化的风景的独特空间里碰面。

③ 这栋建筑式样各异的空间将容纳 150 名员工,2000 名学生。除了几个演讲厅和众多教室外,还创造了致力于创新型教学的空间,还有一个自助餐厅区和休闲区。

④ 建筑物的玻璃正面覆盖有一张宽敞的顶篷,植被涌入内部空间,模糊了内外部界限,对周围环境保持开放、透明和多孔通透。

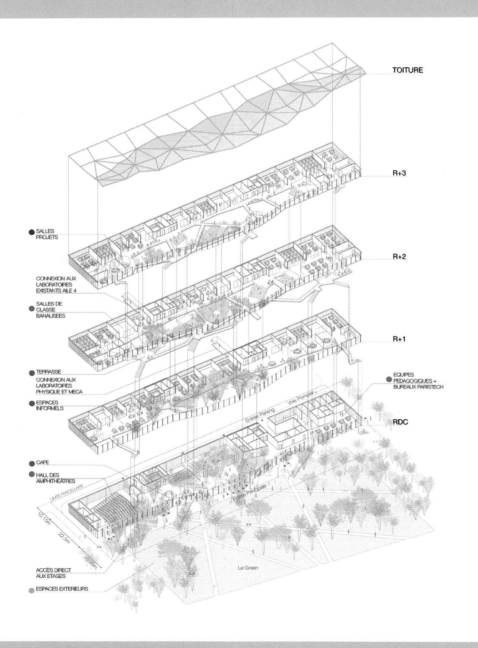

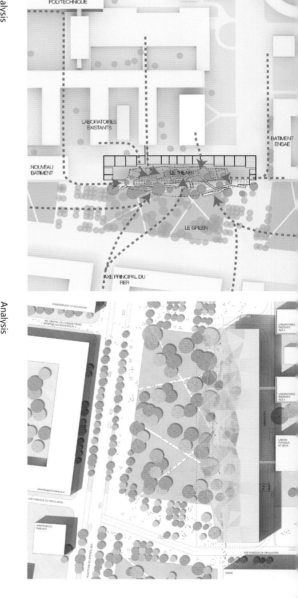

Project Overview / 项目概况

Project overview

The project led by the group Sou Fujimoto, Manal Rachdi Oxo architects and Nicolas Laisné Associates seduced by its openness to the surroundings, its transparency and its porosity. The front glazed facade is covered with a generous canopy while the vegetation is invading the interior space, thus blurring the limits between inside and outside.

1. The video conference and coordination work space like "online learning room", "box work" or "project house" will optimise the environment where interactive and digital tools become the key education system.

2. More informal assembly or work spaces are created so that people may meet in a pleasant and unique space with soft light and alluring landscape instead of on corridors.

3. This building will accommodate 150 employees and 2,000 students. In addition to speech halls and classrooms, it also creates an innovative teaching space, buffet restaurant and entertainment area.

4. The front of the building's glass is covered with a canopy. Plants are introduced into the internal space, blurring its boundary with the outside space, and keeping an open, transparent and ventilated posture to the surroundings.

项目概况

该项目由藤本壮介建筑设计事务所，Manal Rachdi Oxo architects 和 Nicolas Laisné Associates 团队联合领导设计，其对公众的吸引力主要在于它拥有开放、透明的周围环境、以及它的多孔通透性设计。该建筑物玻璃立面覆盖有一个巨大的顶棚，同时绿植布置直接延伸进该建筑物的内部空间，因而模糊了建筑物的内外部界限。

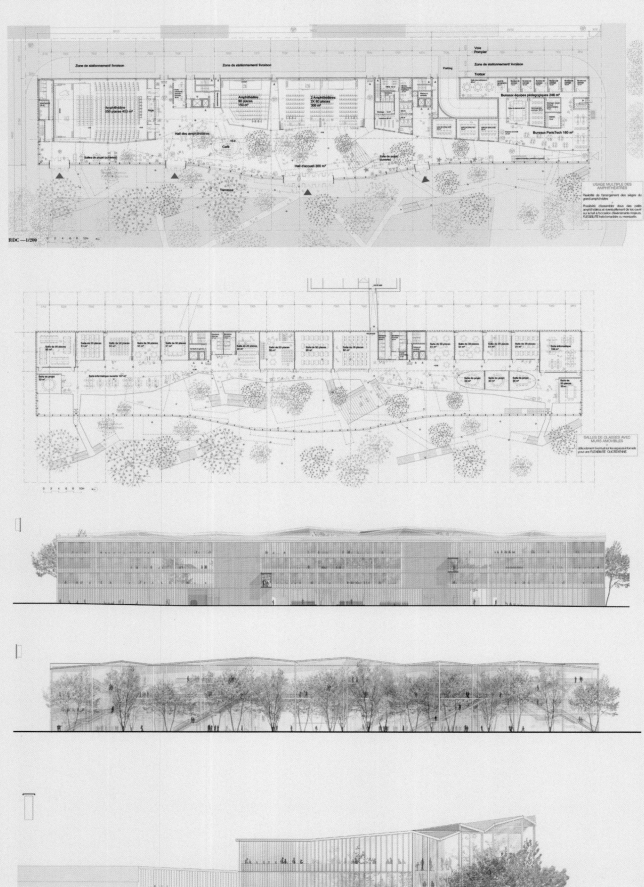

As a symbol of the union of the six schools, all of them belonging to the Paris-Saclay University, the learning centre becomes a point of convergence, thought of as an extension of the qualities of the adjacent landscaping project.

六所学校都隶属于巴黎 - 萨克雷大学，作为一种联合的象征，学习中心成为一个汇聚点，被认为是邻近景观项目特点的延伸。

学习中心将很快在巴黎综合理工学校当前的开发区中心建立起来。这一改革刚进入运行阶段，未来它将是巴黎 - 萨克雷大学城市学校改革的代表。大巴黎快中地铁 18 号线将为校园提供服务，而且校园将达到 174 万平方米的总计划规模。

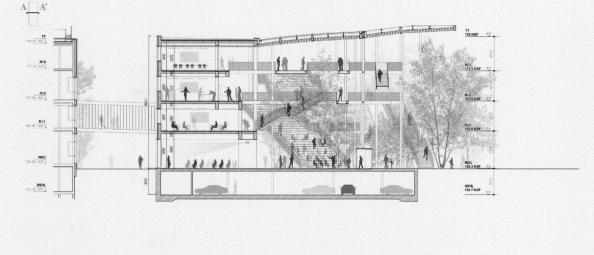

Section AA

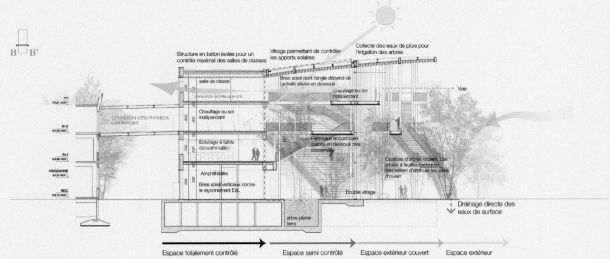

Section BB

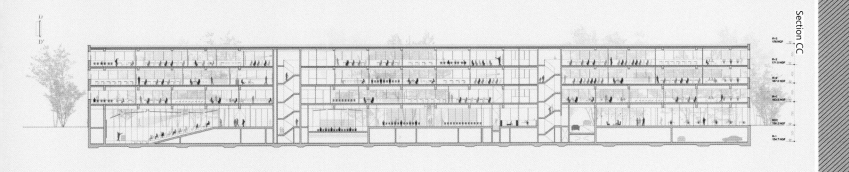

Section CC

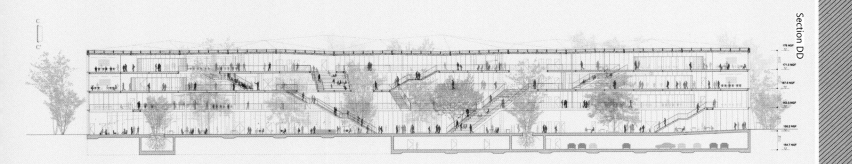

Section DD

The Learning centre, soon to be erected in the heart of the current development area of the École Polytechnique, is representative of the evolution of the urban campus of Paris-Saclay which just entered the operational phase. The campus served by the future train line 18 of the Grand Paris Express will reach a programme of a total of 1.740 million square metres.

Design strategies

The building will host 150 staff and welcome 2,000 students within its premises of various typologies. In addition to several lecture halls and numerous classrooms, spaces dedicated to innovative pedagogy will be created. "Distance-learning rooms", video conferencing and collaborative working spaces such as "boxes-work" or "project rooms" will optimise the environment for an educational system focused on interactivity and digital tools. The building will also provide a cafeteria and relaxation areas favouring serendipity, itself encouraged by the close proximity of the research laboratories of the École Polytechnique.

设计策略

这栋建筑将容纳150名员工以及欢迎2 000名学生入学。除了几个演讲厅和众多教室外,还将创造一些致力于创新型教学的空间。"远程学习房间"、以及诸如"盒子工作"或"项目房"等的视频会议和协作工作空间将会最大地优化这一环境——这一环境中里有一种以互动和数字化工具为重点的教育体系。这栋建筑里同时还设有一家自助餐厅以及能给人们带来意外惊喜的休憩区,因为其紧靠着的就是巴黎综合理工学校的研究实验室,为该建筑增色不少。

Flexibility, mingling and openness

Opening up to the linear park in front, the Learning centre is invaded by nature. Inside, a wide atrium is inhabited by the light vegetation and a series of walkways and staircases creating numerous informal spaces for teachers, students and visitors allowing new places to meet or work. These platforms, the "spontaneous amphitheatres" and the classrooms are united under one roof providing promiscuity and privacy in an intimate relationship with nature. People won't pass each other in corridors anymore, but meet in vivid places, in a unique space bathed in soft light, with surprising and changing views. The large transparent facade of the Learning centre opens to the West on the "Green", a vast public space covered by lawns and partly wooded. The building is thus seen as an open space revealing the activities taking place in its heart and stands as an architectural and academic emblem of the future neighbourhood.

灵活性、混合型和开放性

学习中心向前面的一个带形公园开放，与自然融为一体。在内部，开阔的中庭里生长着明亮的植被，一些走道和楼梯为老师、学生和参观者创造了许多非正式的集会或工作空间。这些平台，"自发形成的梯形座位区"和教室集合设置在一个屋檐下，在与自然的亲密接触中既提供了互动又具有隐私。人们不会再在走廊里擦身而过，而是在生动的场所，在沐浴着柔和光线的、有令人惊讶和不断变化的风景的独特空间里碰面。学习中心宽大的透明外表西面向"绿地"开放，那是一片开阔的铺着草坪、长有一些树木的公共空间。这栋建筑因此被视为是一个开放的空间，展示这在它中心发生的活动，并在未来将成为邻近地区的一个建筑和学术象征。

PROBLEM / 问题

1. **DIFFERENT FROM TRADITIONAL LIBRARY SPACE**
2. **INTEGRATE WITH THE ENVIRONMENT**
3. **FUNCTION**
4. **COMFORTABILITY**

1. 打破传统图书馆空间形式
2. 融入环境
3. 功能性
4. 舒适性

Daegu Gosan Public Library
韩国大邱高山郡公共图书馆

Synthesis Design
works
Synthesis Design
作品

Architect: Synthesis Design
Consultants: Buro Happold LA (Structural/MEP/Facades)
Location: Daegu, Korea
Area: 3,100 m²
Function: Public library

设计公司：Synthesis Design
顾问公司：Buro Happold LA（结构、MEP 以及外墙顾问）
地点：韩国大邱
面积：3 100 平方米
功能：公共图书馆

DESIGN REQUIREMENTS
设 计 要 求

Traditionally, the public library space and social experience are limited to a series of reading rooms separately distributed and reference stacks with defined boundaries and books stacked in disorder, which shares the same challenge in the design of Daegu Goshan Public Library.

人们在传统意义上对公共图书馆的空间和社会体验停留在一系列独立分布的阅读室以及明确规定的界限和杂乱堆叠的书库,这也是设计大邱高山郡公共图书馆所面临的挑战。

① 设计方案提倡一种智能、开放以及综合性的图书馆体验以取代过去的媒体存储方法,并同时借助无所不在的信息资源、集成式室内装饰以及活跃的公共社交空间,将图书馆空间转变为混合环境的综合体。

② 该项目采用上升、下沉以及垂直式多楼层构造的设计手法使得场地好像作为周围公园以及其他城市建筑物延伸出来的一部分。

③ 独特的双表皮设计在避免室内眩光的同时又不会阻隔人观赏室外的视线。

④ 楼板冷热调节系统保证室内温度的舒适度。

Mesh Relaxation Analysis

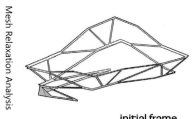

initial frame

simple surfaces

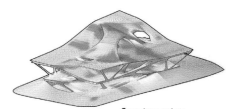

first iteration
mesh relaxation

second iteration
mesh relaxation

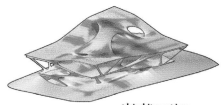

third iteration
mesh relaxation

Axonometric Analysis

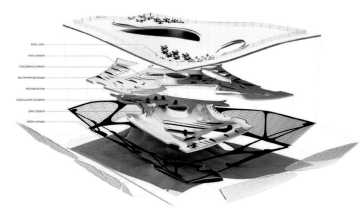

PROGRAM DIAGRAM

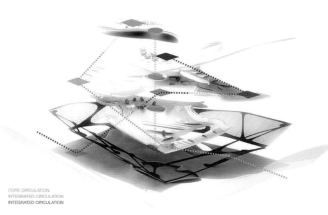

CORE CIRCULATION
INTEGRATED CIRCULATION
INTEGRATED CIRCULATION

CIRCULATION DIAGRAM

Project Overview / 项目概况

Design concept

The proposal for the Daegu Gosan Public Library challenges the conventional understanding of the spatial and social experience of a public library as a series of discrete reading rooms with defined thresholds and cluttered stacks. The architects propose an intelligent, open, and integrated library experience which supersedes the media storage methods of the past

设计理念

人们在传统意义上对公共图书馆的空间和社会体验停留在一系列独立分布的阅读室以及明确规定的界限和杂乱堆叠的书库,这也是设计大邱高山郡公共图书馆所面临的挑战。我们在设计方案中提倡一种智能、开放以及综合性的图书馆体

1 The design encourages a smart, open and comprehensive library experience to replace the original media storage method, and to turn the library into a synthesis with mixed environment through ubiquitous information, collective indoor decoration and active public space.

2 By adopting rising, sinking and vertical storeys design, the project seems like an extra part from the parks and other urban buildings around.

3 The unique double-surface design successfully avoids dazzling indoor while helps you enjoy the views outside.

4 The heat conditioning system ensures the temperature comfortability indoor.

Structure Analysis

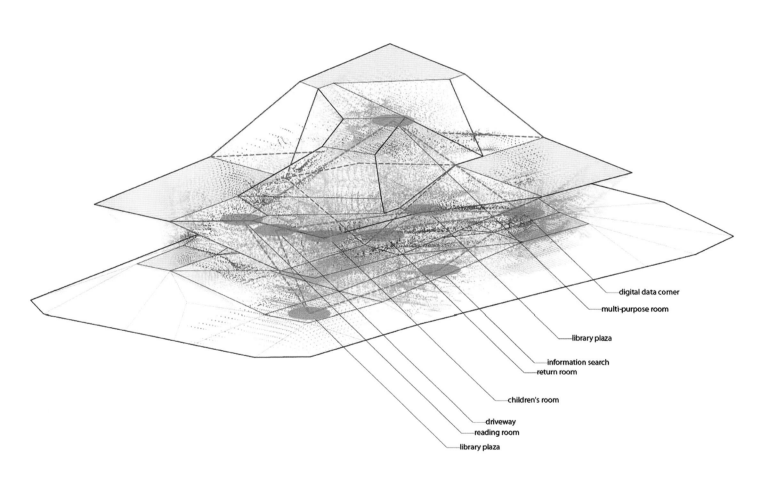

- digital data corner
- multi-purpose room
- library plaza
- information search
- return room
- children's room
- driveway
- reading room
- library plaza

and changes the library space into a hybrid environment through ubiquitous information resources, integrated furnishings and active communal social spaces. The architecture is designed to enable and embody the spirit of open-source exchange and collective knowledge through free-form geometries, open plans and integrated amenities. The architects have minimised the thresholds between spaces, floors, and functions to consider the library as an active, continuous, and fluid field of social, cultural, and intellectual discourse.

验以取代过去的媒体存储方法，并同时借助无所不在的信息资源、集成式室内装饰以及活跃的公共社交空间，将图书馆空间转变为混合环境的综合体。凭借自由形式的几何图形设计、开放式的布局以及配备的综合便利设施，该项目建筑大楼被设计成能激发并展现开放性资源交流以及集体知识分享精神的空间。我们已经最小化不同空间、不同楼层以及不同功能区之间的界限，力争将整个图书馆设计成一片活跃、连续而流畅的区域，涵盖一系列社会、文化以及知识著述。

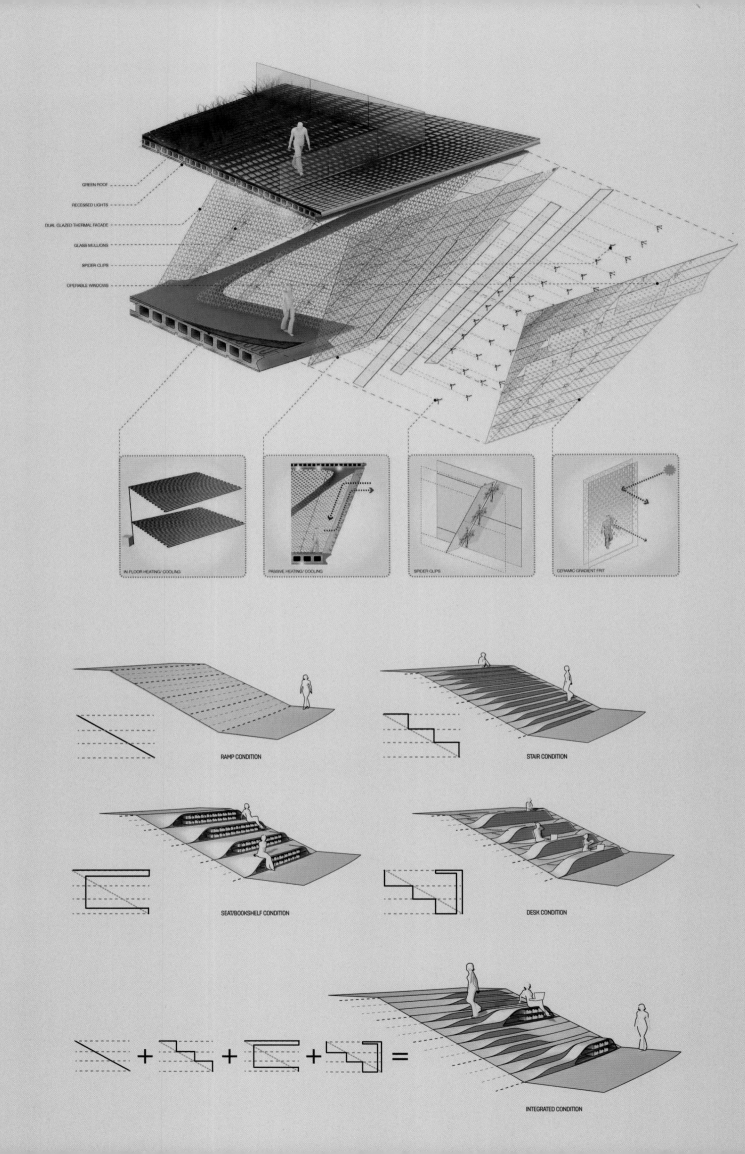

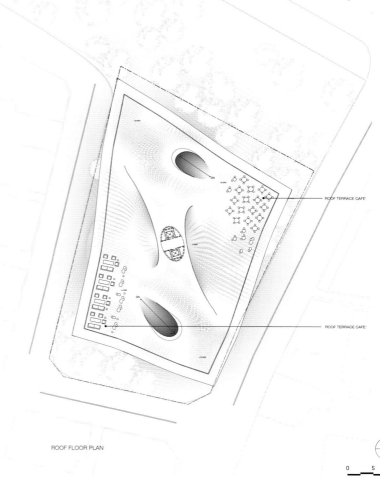

3RD FLOOR PLAN

ROOF FLOOR PLAN

Floor Plan

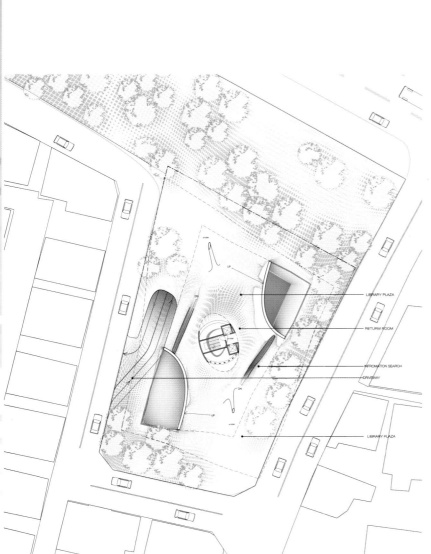

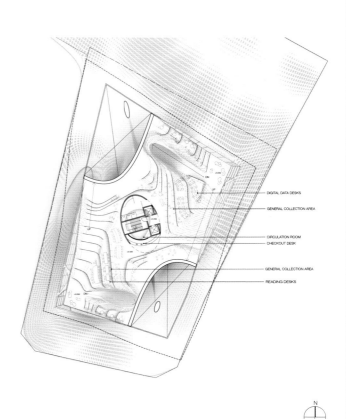

GROUND FLOOR PLAN

2ND FLOOR PLAN

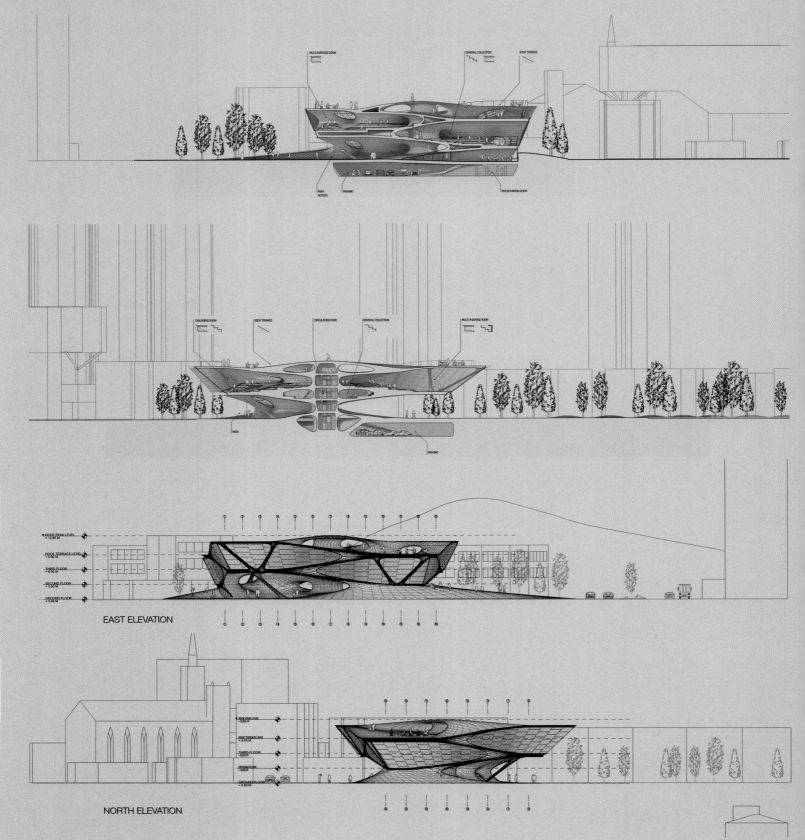

Design strategies

Conceptually and literally, the ground field of the site swells, peels, and multiplies vertically as a continuation of the adjacent park and urban fabric. This constant sectional change is articulated as a smooth vertical gradient which merges floors, ramps, stairs, terraces and furnishings into an inhabitable and ergonomic landscape culminating in an open-air roof-scape lounge and terrace overlooking the city of Daegu. The boundaries between floors are blurred, as the continuously walkable surface which unifies the many spaces of the library facilitates circulatory, physical and visual connections both internally within the network of spaces and externally with the surrounding context.

The building is materialized as an in-situ reinforced concrete structure which, like all other aspects, has been designed to be fully integrated with the geometry of the library. The central core of the building provides its primary structural point of reference connecting vertically through the entire structure. The free-form geometry that defines the walkable surfaces and unifies the building cantilevers out from this central core and is supported by its internal interconnections (ramps), perimeter interconnections (columns) and the lifted ground plane

设计策略

不管是在理论上还是在实践中,该项目场地作为周围公园以及其他城市建筑物延伸出来的一部分,采用的设计手法包括上升、下沉以及垂直式多楼层构造。该项目大楼的组合方式处于不断的变化之中,并最终演变成流畅的垂直梯度式分布方式,将各楼层、坡道、楼梯、露台以及室内装饰全部融入进这个适合居住而又采用人体工学设计的景观之内。该景观最大的特色是其包括一个可欣赏屋顶风景的户外休息厅和一个可俯瞰整个大邱城市风景的露台。各楼层之间的界限并不十分清晰,因为适于人们通行的连续性曲面设计可将图书馆的诸多空间统一起来,从而加强该图书馆大楼内部空间网络之间循环、物理以及视觉连接以及其与外部周围环境融合。

该图书馆大楼实质上就是一个就地浇筑的钢筋混凝土结构——像所有其他方面一样,该结构已完全和图书馆的几何构造融合为一体。该大楼的中心位置起着主要参考结构点的作用,在垂直方向将整个结构贯穿连接起来。自由形式的几何

(foundation). The geometric logic of the form has been developed through a computational method known as "dynamic mesh relaxation" which relaxes planar mesh networks to "form-find" a continuously minimal surface. As developed in the 1950's and 60's through the seminal work of Frei Otto, minimal surfaces articulate the natural force paths of structural loads thus providing the optimal shape for maximum structural performance with minimal material. The shape of the surface, thus allows for relatively thin structure, which in this case is materialised as cast-in-place high performance reinforced concrete. This concrete would be cast on CNC-milled EPS foam formwork, coated with polyurethane. The geometry of the building has been rationalised so that each piece of formwork could be reused at least four times in order to maximise the efficiency and economy of the process.

构造体明确划分了其可供人通行的曲面，并将从中心区域延伸出来的大楼悬臂式设计部分连接起来，而大楼内部的互联点（坡道）、外缘互联点（立柱）以及抬高的水平面（地基）将为该几何构造体提供支撑力量。通过一种被称为"动态网格松弛"的计算方法，该建筑形式的几何逻辑性思维便被开发了出来。该方法是通过松弛平面网状网络结构来为一个最小的连续性曲面"找型"。得益于弗雷·奥托大师的开创性设计工作，该方法在1950至1960年代得到了长足的发展。最小曲面可清晰地显示出结构荷载的自然受力路径，因而可在使用最少材料的情况下，为实现最大结构性能设计出最优的形状。我们可根据曲面的形状采用相对较薄的结构设计。因而，在该项目的设计上，我们选用的是就地浇筑的高性能钢筋混凝土。钢筋混凝土将被灌注在经CNC研磨的EPS泡沫模子内，表面覆有一层聚氨酯涂层。该图书馆大楼的几何构造体形状都比较规范标准。如此一来，每一块模子都至少能够被重复利用四次，以达到效率最大化以及节约建设成本的目的。

PROBLEM / 问题

1. **AUDIO VISUAL SYSTEM**
2. **FUNCTION**
3. **ENERGY-SAVING**

1. 视听系统
2. 功能性
3. 节能性

House of Music, Aalborg, Denmark
丹麦奥尔堡音乐厅

COOP HIMMELB(L)AU
works
COOP HIMMELB(L)AU
作品

Architect: COOP HIMMELB(L)AU
Client: North Jutland House of Music Foundation, Aalborg, Denmark
Location: Aalborg, Denmark
Net Floor Area: 17,637 m²
Gross floor area: 20,257 m²
Function: Concert hall

设计公司：COOP HIMMELB(L)AU
项目客户：丹麦奥尔堡日德兰半岛北部音乐厅基金会
项目地点：丹麦奥尔堡
净建筑面积：17 637 平方米
总建筑面积：20 257 平方米
功能：音乐厅

DESIGN REQUIREMENTS

设 计 要 求

This project is house of symphony. The houses of music, education and performance are aimed to provide communication information, knowledge and inspiration for local residents and those using the music houses.

本项目为交响音乐厅，音乐厅、教育厅和表演厅，旨在为当地居民和使用音乐屋的人们提供交流信息、知识和灵感的地方。

① 该项目建筑公司与 Arup 工程的 Tateo Nakajim 合作，设计出了一套高度复杂的音响系统以确保观众可获得最优的视听体验。

② 紧凑的 U 形地块中心是 1 300 座的世界一流的交响音乐厅，音乐厅、教育厅和表演厅环绕在交响音乐厅三面。

③ 通过设计自然通风热交换，安装热交换器设备和低风速高效率的通风系统等措施减少不必要的能源消耗。

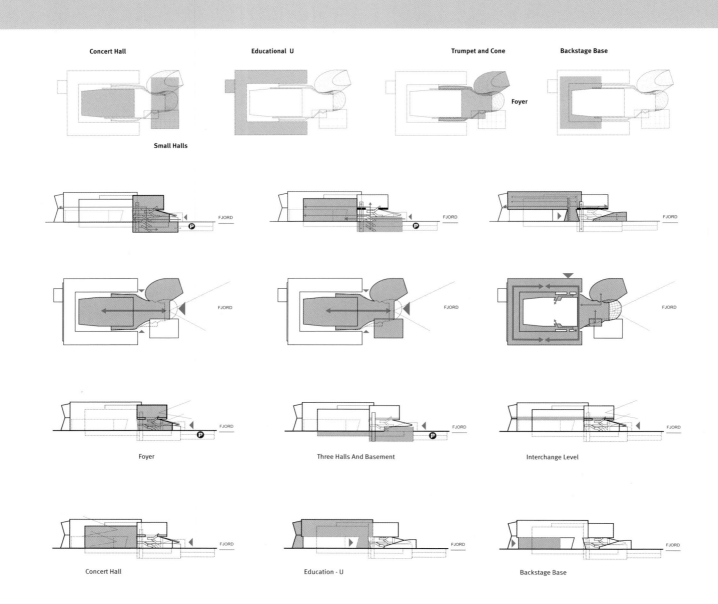

Project Overview / 项目概况

Location analysis

U-shaped rehearsal and training rooms are arranged around the core of the ensemble, a concert hall for about 1,300 visitors. A generous foyer connects these spaces and opens out with a multi-storey window area onto an adjacent cultural space and a fjord.

The concert hall

The seats in the orchestra and curved balconies are arranged in such a way that offers the best possible acoustics and views of the stage. The highly complex acoustic concept was

区位分析

整个项目地块的中心是一座世界一流的，可以同时容纳1300位观众的交响音乐厅，四周被多个U形的排练厅和培训室环绕着。建筑向北面打开，形成一个宽敞的公共休息区域，将这三个空间巧妙地连接了起来，并缓缓地向几层楼高的观察窗区域展开，可俯瞰相邻的文化广场和峡湾。

1. The project construction company cooperates with Tateo Nakajim of Arup project to design a set of highly complex audio system so that the audience may enjoy the best audio visual experience.

2. The centre of this compact U-shaped land block houses a world-class house of symphony, with houses of music, education and performance surrounding three sides of the house of symphony.

3. The natural ventilation system, heat exchanging equipment and low-speed high-efficiency ventilation system are designed to reduce unnecessary energy consumption.

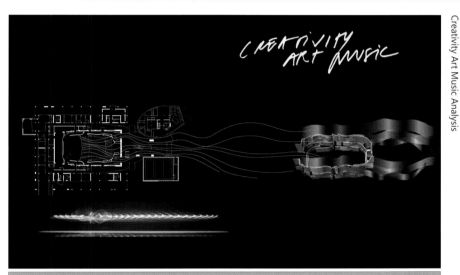

Creativity Art Music Analysis

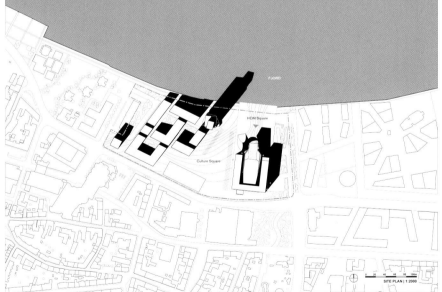

Site Plan

developed in collaboration with Tateo Nakajima at Arup. The design of the amorphous plaster structures on the walls and the height-adjustable ceiling suspensions, based on the exact calculations of the specialist in acoustics, ensures for the optimal listening experience. The concert hall will be one of the quietest spaces for symphonic music in Europe, with a noise-level reduction of NR10 (GK10).

音乐厅

交响音乐厅内座椅区的特别排列以及曲线状的阳台设计则帮助创造出最佳的音响与视野效果。实际上，该项目建筑公司与 Arup 工程的 Tateo Nakajim 合作，设计出了这一套高度复杂的音响系统。音乐厅内部墙面上的不规则石膏板结构以及高度可自由调节的天花板悬挂装饰物是根据声学专家们的精确计算而专门设计出来的，以确保观众可获得最优的视听体验。(如何提高视听体验) 该音乐厅建成之后，将成为整个欧洲噪音程度最小的交响乐音乐厅之一，噪音程度降低至 NR10 (GK10)。

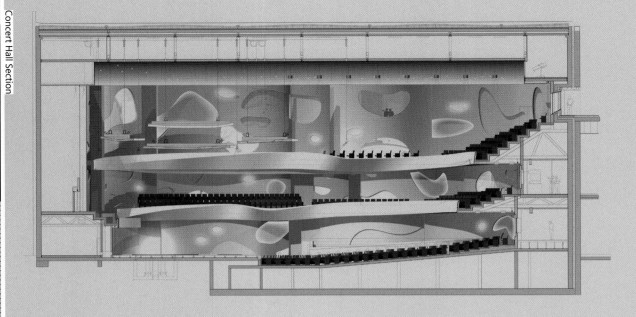

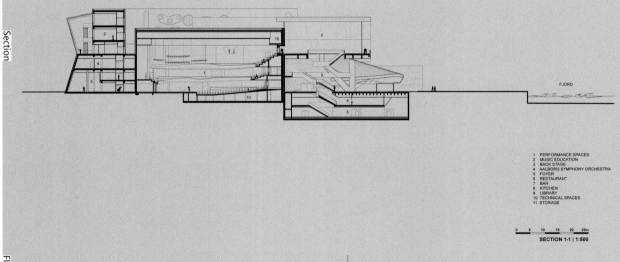

SECTION 1-1 | 1:500

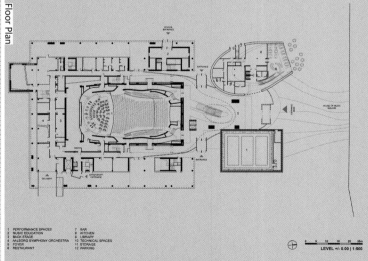

LEVEL +/- 0.00 | 1:500

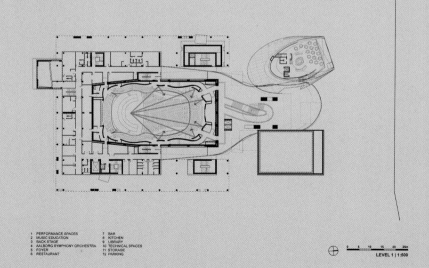

LEVEL 1 | 1:500

The foyer
The foyer serves as a meeting place for students, artists, teachers, and visitors. Five storeys high with stairs, observation balconies, and large windows with views of the fjord, it is a lively, dynamic space that can be used for a wide variety of activities.

Other three halls
Under the foyer, three more rooms of various sizes complement the space: the intimate hall, the rhythmic hall, and the classic hall. Through multiple observation windows, students and visitors can look into the concert hall from the foyer and the practice rooms and experience the musical events, including concerts and rehearsals.

公共休息大厅
公共休息大厅是一个能够汇聚学生、艺术家、老师和游客的聚会场所。五层楼高的大厅内阶梯盘旋穿梭，有着大型观景窗的阳台可一览峡湾美景。大厅如同是个动态十足的空间，随时可变化成举行各类活动的场所。

其他三个小厅
在公共休息大厅的下面是三个额外的小厅（即私人大厅、韵律馆以及古典馆），它们的空间大小不一，功能与交响音乐厅互补。借助几层楼高的观察窗口，学生们和参观人员可从公共休息大厅以及练习室中一览音乐厅的风采，并亲身感受各种有趣的音乐节目表演，包括音乐会和彩排等。

Energy efficiency design

Instead of fans, the foyer uses the natural thermal buoyancy in the large vertical space for ventilation. Water-filled hypocaust pipes in the concrete floor slab are used for cooling in summer and heating in winter. The concrete walls around the concert hall act as an additional storage capacity for thermal energy. The fjord is also used for cost-free cooling.

The piping and air vents are equipped with highly efficient rotating heat exchangers. Very efficient ventilation systems with low air velocities are attached under the seats in the concert hall. Air is extracted through a ceiling grid above the lighting system so that any heat produced does not cause a rise in the temperature in the room. The building is equipped with a building management programme that controls the equipment in the building and ensures that no system is active when there is no need for it. In this way, energy consumption is minimised.

节能设计

在不使用风扇的情况下，利用公共休息大厅垂直的高度来进行自然通风热交换。混凝土地板下的水管用来作为温度系统调节使用（夏天降温，冬天则保温）。而四周的混凝土墙成了额外的储存热能的体量，一旁的峡湾则提供了天然的冷却环境。管道系统和通风系统都配备了高效率旋转式的热交换器设备。同时，音乐厅每一个观众席位下面还安装有低风速高效率的通风系统。在天花板网格的照明系统上方设置了抽风系统以降低因上升热空气所导致的上升的室内温度。而建筑内的中央控制系统则确保设备在不需使用时不被启动，以减少不必要的能源消耗。

PROBLEM / 问题

1. **CREATING A LANDMARK SHOWING THE HUNGARIAN CULTURE**
2. **RESPECTING THE HISTORICAL CITY PARK**
3. **CREATING OPEN, TRANSPARENT, ALLURING COMMUNITY SPACES FOR THOSE LIVING IN BUDAPEST**
4. **CREATING A LASTING AESTHETIC EXPERIENCE TO THE VISITORS**

1. 创建具有匈牙利文化内涵的地标建筑
2. 尊重历史城市公园气质
3. 为布达佩斯居民创造一个开放、透明、迷人的社区空间
4. 让游客感受永久性美的体验

Museum of Ethnography
民族志博物馆

BFarchitecture
works

BFarchitecture
作品

Architects: BFarchitecture
Team: Eva Trip, Friso Jonker
Location: Budapest, Hungary
Area: 20 000 m²
Function: Museum

设计公司：BFarchitecture
团队成员：Eva Trip，Friso Jonker
地点：匈牙利布达佩斯
面积：20 000 平方米
功能：博物馆

① 在一个盒子体积上，弯弯曲曲创建一个独具匈牙利风格的木质结构。

② 通过巧妙运用高密度都市和公园以及室内外空间的过渡，博物馆的外形借鉴了阴阳之说。

③ 民族博物馆的设计灵感来自于一个公共市场或集会

④ 商铺、咖啡馆和运动空间之间由一条走廊直接连接。一条宽大的斜坡将带领游客从接待处走向博物馆不同区域。

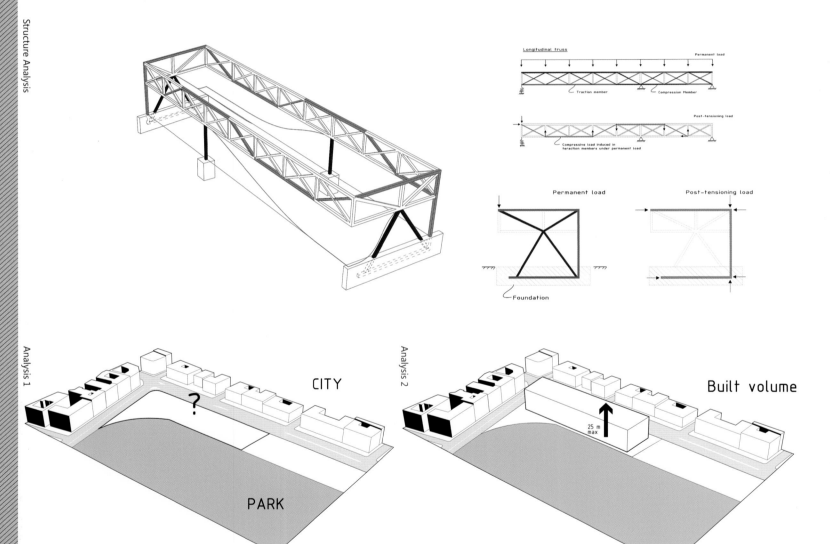

The location of the site for the Museum of Ethnographie is seen as a link between the city and the park. How can we imagine The Museum of the Peaple as social experience through its unusual context?

Basic volume with a maximum hight of 25m on the site.

Project Overview / 项目概况

Project overview

The Museum of Ethnography brings together three cultural institutions: Museum Main Exhibition "temporary and permanent"/ Museum Learning/ Museum for Children and their shared facilities.

Creating a layer made of wooden scales pieces with typical Hungarian style, bows on a twisted movement in a box volume.	①
The shape of the museum refers to yin and yang by forming a transition between the dense city and the park and between outdoor and indoor spaces.	②
The design of the Museum of Ethnology is inspired by a public market or agora.	③
A promenade starts from shops, café to events spaces. A large ramp takes the visitors from reception to the different ambiance.	④

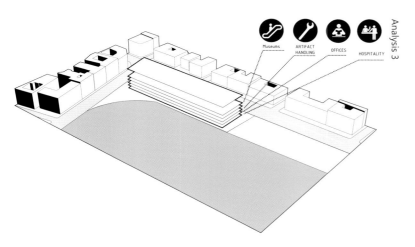

The Built volume turned on layers combining three cultural institutions (Museum Exhibition/ Museum Learning/ Museum for Children) and their shared-facilities.

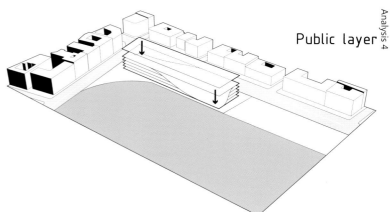

the three institutions (Museum Exhibition/ Museum Learning/ Museum for Children) and their facilities are gathered arround a public space: An Outdoor Urban Layer open towards both the city and the park.

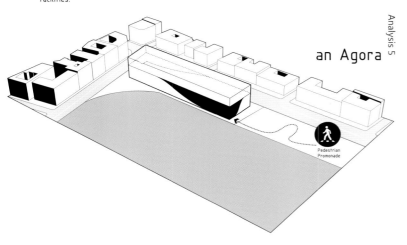

The public layer- an Agora will reflect the evryday life of the city flowing through its generous terraces and staircase along the promenade from Dózsa György út.

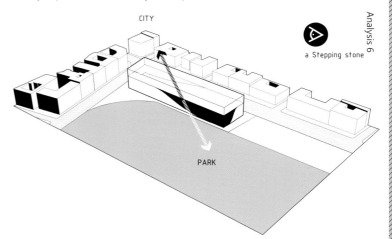

On sepecial occasions it will turn into an urban stage or an outdoor gallery to extend art into the city as well as city into the architecture.

项目概况

民族志博物馆是由三大文化机构合并而成,即主展览博物馆,包括临时展览和常设展览;学术博物馆;儿童博物馆及相关公用设施。

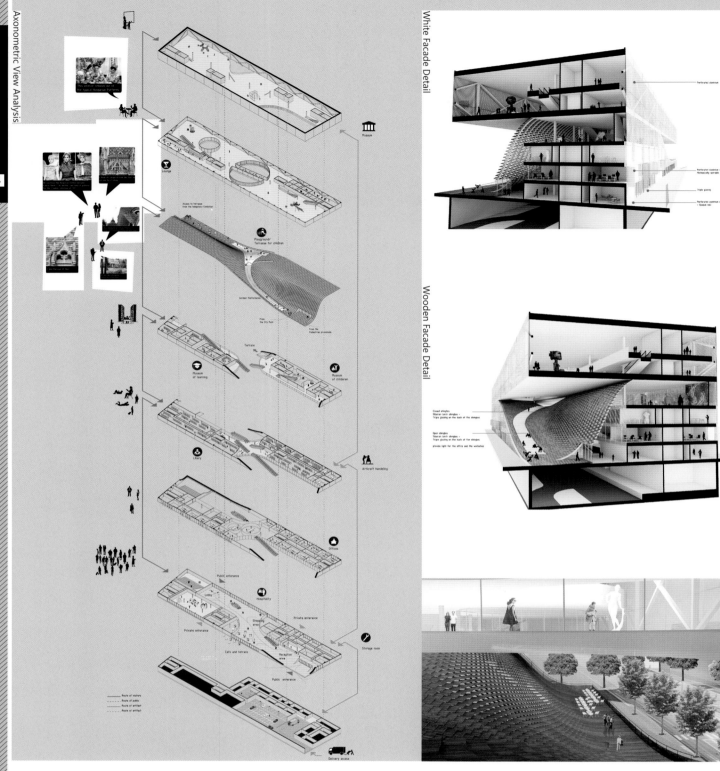

Design strategies
Contrast

The contrast between the dense city and the park articulates a true urban experience, the proposal offers a building that is intertwined with the park and the city. The three institutions (Museum Exhibition/ Museum Learning/ Museum for Children) and their facilities are gathered around a public space – an outdoor urban layer. The building layers and the outdoor urban layer flow together in a mutual embrace towards both the city and the park. Almost as yin yang symbol, the concept is easily perceivable from the outside.

Connection

The Museum of Ethnography in Budapest is conceived as a confluence of the park and the city – nature and architecture. Bookending the promenade along Dózsa György út, the building allow public life to follow through it. The urban pavement, starting from The Museum of Fine Art and passing by the Hero Square, rises to become a pocket of terrace overlooking the park and elevating an outdoor gallery above the city. As an Agora, the museum of the people will reflect the everyday life of the city flowing through its generous terrace and staircase

设计策略
对比

密集的城市与公园之间的对比清晰地表达了一种真正的城市体验。在我们的设计提案中，该项目建筑将与公园和城市交织融合于一体。三大文化机构（展览博物馆，学术博物馆以及儿童博物馆）以及它们的配套设施将集中设置在一个公共空间内——即一个户外城市层。该项目大楼的楼层与户外城市层相互拥抱连接于一体，在整体上与整个城市和公园交相呼应。就跟中国传统文化中的"阴阳"标志一样，（从远处观看）该项目建筑的设计理念是很容易被人们理解的。

联系

布达佩斯民族志博物馆被设计成为一个公园与城市的统一体，达到自然与建筑物完美融合的效果。该项目建筑邻近Dózsa György út旁边的人行道，融入了浓浓的生活气息。这条城市人行道，起点位于美术博物馆，然后经过英雄广场，

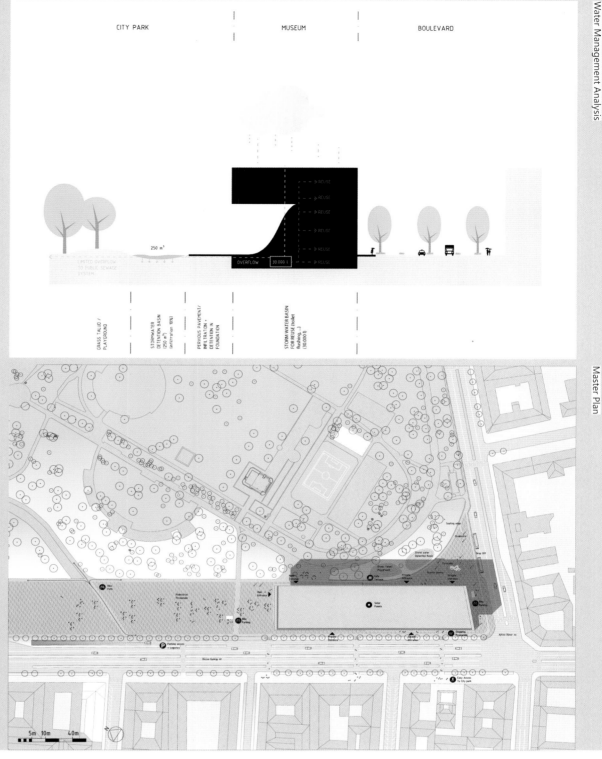

Water Management Analysis

Master Plan

where people, functions and activities meet to form a dynamic and vibrant space. On special occasions it will turn into an urban stage for an outdoor performance to extend the art into the city as well as the city into the architecture.

Movement

A space where the users articulate the building while the building articulates the city. On the ground floor, the promenade continues on the inner attractive path from shops, café to events spaces. A large ramps take its starting point at the busy arrival reception and gradually extends to different ambiances. The Museum becomes more than exhibition space; it's a space to be.explored and experienced. The ramp leads visitors to cross the institution part of the museum with all its workshops, ethnological archives and library to the main exhibition part. The exhibition core is lifted above the institutional pillars. From the Main exhibition, a lounge bar is taking a place as a relaxing point after the exhibition visit. The lounge bar is extended on the terrace above the city. An outer staircase leads the visitors to rejoin the Park side and the Pedestrian promenade.

逐渐上升成为呈台阶状分布的围合之地，可俯瞰公园的风景，并成为整个城市之上的一条户外走廊。人民博物馆是一个城市广场，开放式的露台和楼梯上每天人来人往，生活气息相当浓厚。人们相聚于此，利用该城市广场的各种功能举行着各种活动，俨然已将人民博物馆变成一个活力四射的动态空间。在特殊的节日里，它将转变成人们举办户外表演的城市舞台，将艺术融入城市，同时也将城市融入建筑物。

运动

在这片空间里，使用者清晰地定义了该建筑物的功能，而该建筑物却反映了整个城市的氛围。在建筑大楼的第一层，人行走廊延伸了大楼内部有趣的线路布局，并在这一整条沿线路径上将商店、咖啡馆和娱乐活动空间串联起来。一条大型坡道，起点始于繁忙的下客接待中心，并逐渐地向不同的周边环境延伸。该民族志博物馆不仅仅是展览空间，它还值得人们去不断探索，获得不同体验。游客们经由这条坡道可以横穿博物馆的三大文化机构区域，并途径各大工作间、民族志档案室和图书馆，最后到达主展览区域。展览的核心区域设置于文化机构区域的立柱之上。从主展览区域出来之后，游客们可在看完展览后到酒吧稍作休憩。酒吧空间一直延伸至露台，可俯瞰整座城市的夜景。游客们可经由建筑大楼外部的楼梯重新到达公园区域以及人行走道区。

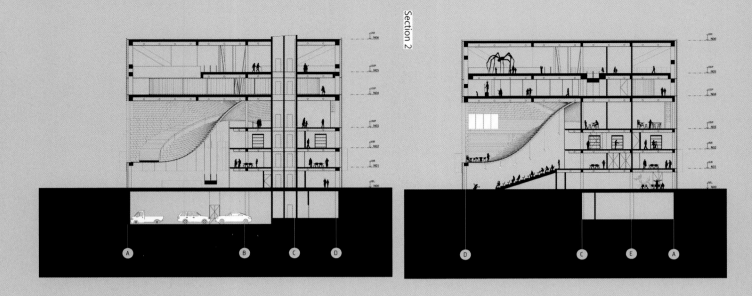

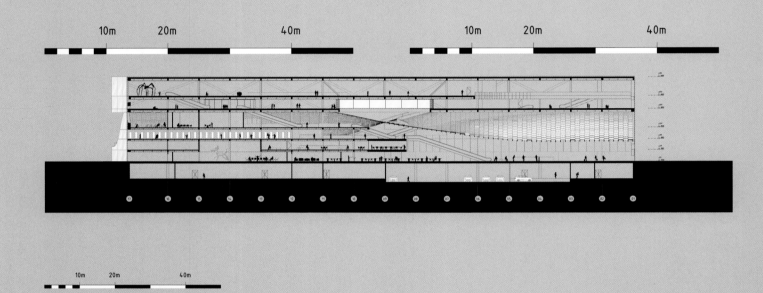

Design points

1. Traffic design

In order to enhance the relation between the city centre and the City Park, the Dòsza Györgi út requires a redesign that seperates local from through traffic. Several green resting points improve the crossability for pedestrians and bicycles drastically. The redesign of a small part of the park creates a small entrance square that continues as well on the wave of the building, as to the interior entrance hall. Covered bicycle parkings with integrated charging stations for electrical bikes are situated on the main entrances of the museum. Public transport, the trolley bus, can have an extra stop alongside the terrace of the museum and be combined with a drop-off bay for three touristic buses. The underground parking under the agora has the intrinsical quality of linking the three museums along the Dòsza Györgi út. Through a smart design of the upper parking layers, the logistic services of the three museums would be facilitated underground instead of by separate ramps that divide the public space. A link between the underground public parking and the private parking spaces for the museum can be made as well on this level. Some of the private parking spaces are equipped with charging stations.

设计要点

1. 交通设计

为了加强城市中心与该城市公园的联系，Dòsza Györgi út 要求对公园布局重新进行设计，完善该区域的交通状况。通过设置几个绿色休憩点，该区域的人行道与自行车道的交叉分布情况得到了较大改善。城市公园的一小部分经重新设计，形成了一个小型的入口广场——随着整栋建筑大楼波动起伏的轮廓节奏，可直达博物馆的内部入口大厅。博物馆的主入口区域上设有多个自行车停靠点，以及专门为电动自行车设置的综合充电站。公交车和无轨电车可沿着博物馆的露台区在停靠一站，同时还专门为开往博物馆的旅游巴士设置了一站下客区。城市广场下面的地下停车场本身就可以起到过渡连接的作用，可将沿着 Dòsza Györgi út 分布的三个博物馆串联起来。通过巧妙设计停车场上面的层次，可进一步完善三大博物馆区的地下物流服务，而无需划分公共空间，设置多条独立的坡道。在这个层面上，同时也可以将地下公共停车场与博物馆私人停车空间连接起来。而且，有一些私人停车空间还专门设有充电站点。

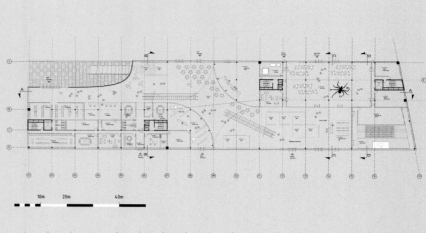

Ground Floor Plan

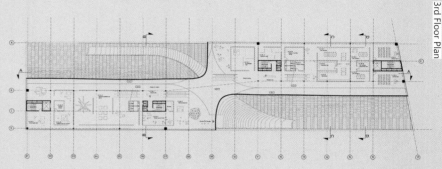

3rd Floor Plan

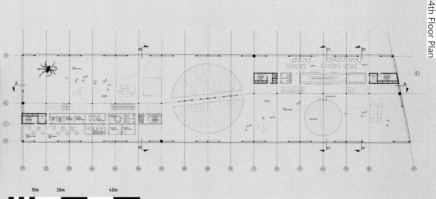

4th Floor Plan

2. Water management

Water for reuse is contained in a water basin in the basement. This way the storm water can be used for toilet flushing and watering plants. The pavement on the parking roof can have a water buffering system as foundation. The overflow of reuse tank and the water buffer under the pavement, flows into a natural detention pond that allows infiltration, but provides a nice natural water element with natural banks as well and guarantees a supply of firewater in case of emergency.

3. Structural design

The intended architectural expression for the exposition space on levels 4 and 5 is that of a simple closed box placed down on a curve-shaped surface. The curved surface creates a vast open space under this box and creates a diagonal passage allowing to cross from one corner of the building to another, while remaining sheltered by the top structure. Structurally, the limited number of supports and the important spans up to 88m are resolved by a 10m high steel truss in each facade. Longitudinally the static schema of the truss can be simplified to a beam supported at three points. Two of these are direct supports on the columns of the lower levels. The third support at the end of the truss is situated above the void created by the curve-shaped surface. Its reaction is taken up by a cantilever structure on the short facade.

2. 用水管理：

可重复利用的水都储存在地下室的水池中。如此一来，暴风雨天气下的雨水可以收集起来用于冲厕所和给植物浇水。停车场上方的人行道可以安装一个水缓冲系统作为地基。从循环用水储水池以及人行道水缓冲区溢出的水将进入一个天然的储水池塘，水会自动往下渗，成为补充天然河道的绝佳水资源，而且在发生突发火灾情况时，可作用为灭火用水。

3. 结构设计：

位于第四和第五层的展览空间的建筑表现形式按照原设计计划将被设置成一个简约的、放在弧形表面下方的闭合箱式空间。弧形表面可在该箱式空间的下方创造一个巨大的开放式空间并形成一条沿对角线设置的走道，以便在同一顶部结构的庇护下，直接从建筑大楼的一角直接横跨到达另一角。从结构上来说，通过在每一个立面内部使用一个10米高的钢桁架，有限的支承数量以及长达88米的跨距问题都迎刃而解。在水平方向上，钢桁架在静态模式下可以精简为通过三点进行支撑的横梁，其中的两个支撑点直接作用在较低楼层的立柱上，而钢桁架末端的第三个支点则设置在由弧形表面形成的空缺处。它的反作用力则由较短外墙的悬臂结构承受。

4. Facade and envelope

The "white facade" consists of a stick system triple glazed curtain wall from floor to ceiling. In front of the glass unit an external shading system with perforated aluminium sheet is installed with motorised movable panels (Folding: pivoting-sliding) for excellent view out, high performance shading and maintenance purposes. The inner layer of the facade is a triple-glazed insulated high performance glass wall to achieve optimum energy and comfort performances in the office spaces. Due to the high g-value of the transparent parts of the facade in

4. 立面与封层

"白色立面"是由从天花板一直垂到地板的三层釉面幕墙组成。在这一玻璃装置的前方，安装的是多孔铝板外部遮光系统，该系统采用了机械化可移动面板设计（可通过绕轴旋转并滑行移动的方式将其折叠起来）以便欣赏外面美丽的风景，同时达到高效遮阳以及高效维护的效果。

外立面的里层采用的是三重釉面高效能隔热玻璃墙，不仅可以使办公空间达到最优的能耗性能，而且还能提升整个空间的舒适度。由于外立面透明部件的G值'很高'，再配备有可移动式的外部遮光装置，在晴朗的冬季天气里，可最大

combination with movable external shading devices, a maximum use is considered of the passive solar heat gain during colder but sunny periods. This highly insulated wall will also reduce dramatically the asymmetric radiation and thus improve thermal comfort in the rooms. The perforations of the external skin of the facade are designed and optimised with daylight and energy simulation tools; resulting in different perforation rates per orientation controlling the solar heat gains, daylight entrance and view-outs as necessary: 63% NE facade towards the park; 54% SE and NW facade; 40% for the SW facade facing the city.

化使用被动式日照得热量。具隔热性能极高的墙面同时也可以大大降低太阳光的非对称辐射，并因此提高各个房间的热舒适度。
外立面的外表皮层采用的是多孔设计，并将根据日光和能量仿真工具的情况调整到最优化使用状态；为了在必要的时候对太阳热增益，日光摄入量以及窗外可视程度进行管控，每个朝向的穿孔率也是采用不同设置的：朝向公园的东北外立面的穿孔率为63%；东南以及西北外立面为54%；朝向城市的西南外立面为40%。

PROBLEM / 问题

1. **FLEXIBILITY**
2. **OPENNESS**
3. **FUNCTION**
4. **ENERGY-SAVING**

1. 灵活性
2. 开放性
3. 功能性
4. 节能性

New Taipei Museum of Art, China

中国新台北市艺术博物馆

DCPParquitectos works

DCPParquitectos 作品

Architect: DCPParquitectos
Location: Taipei City, Taiwan province, China
Function: Museum

设计公司：DCPParquitectos
地点：中国台湾省台北市
功能：博物馆

DESIGN REQUIREMENTS
设计要求

The design is expected to create an open and welcoming space, which can remove the communication barriers among artists, sponsors, experts and the public, eliminating the influences of exclusiveness.

设计初衷是希望将其设计成一个开放式、欢迎人们造访的空间,并消除艺术家、资助人以及专家与普通大众之间的隔阂,破除排他性主义的影响。

① 博物馆采用立柱骨架的设计，给予博物馆管理者充足的空间来设计安排任何独特需求的各种展览和顺序。

② 林林总总的立柱一直从博物馆外部空间延伸至内部空间，模糊了内外空间的界限，营造出一种独特的朦胧美。

③ 独特的表皮设计以抵挡直射光的射入，满足室内特殊的采光需求。

④ 采用雨水收集系统和新风系统等被动式节能技术，以保证室内的恒温。

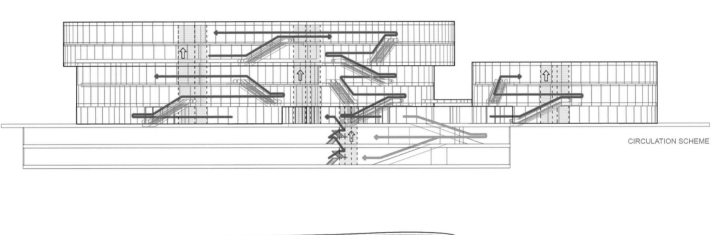

CIRCULATION SCHEME

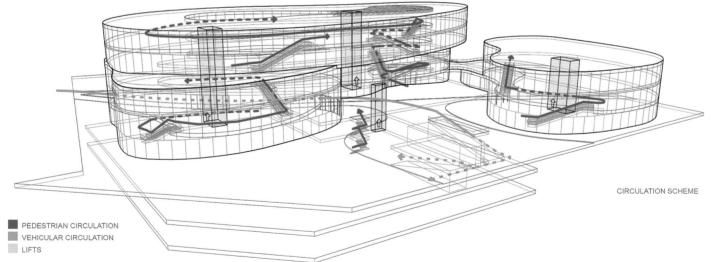

- PEDESTRIAN CIRCULATION
- VEHICULAR CIRCULATION
- LIFTS

CIRCULATION SCHEME

Project Overview / 项目概况

Design concept
The proposal for the New Taipei City Museum of Art is an open and welcoming design that erases the barrier of exclusivity normally surrounding the world of art, patrons, and experts. As such, the architecture of the New Taipei City Museum of Art is one that embodies this idea of erasure through eliminating the traditional borders between exhibition space and

设计理念
我们对新台北市艺术博物馆的设计方案是将其设计成一个开放式、欢迎人们造访的空间，并消除艺术家、资助人以及专家与普通大众之间的隔阂，破除排他性主义的影响。

The museum adopts upright posts as its framework, allowing the management to freely design various exhibition halls with any different demands and sequences.	❶
Various upright posts are extending from the outside into the inside space, blurring the boundary between spaces, and creating a special hazy beauty.	❷
The unique surface design helps withstand the direct sunlight, but satisfy the special lighting demands indoors.	❸
Passive energy-saving technologies like rainfall collecting system and air system are used to ensure a constant indoor temperature.	❹

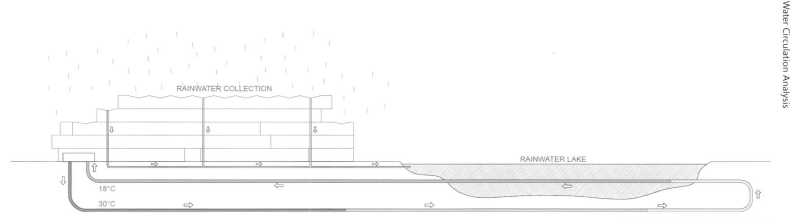

Water Circulation Analysis

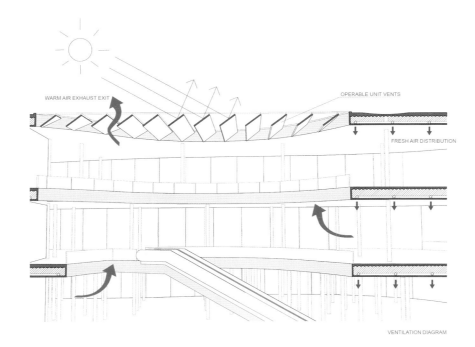

Ventilation Analysis

circulation, as well as exterior and interior. Every part of the museum is represented by a space without limits that can hold any type of expression.

如此一来,新台北市艺术博物馆大楼将成为"消除隔阂"设计理念的一个代表作——消除展览空间与动线之间存在的传统意义上的界限,以及大楼内外部之间的界限。博物馆的每一部分都代表着无界限的一个空间,可容纳任何形式的表达。

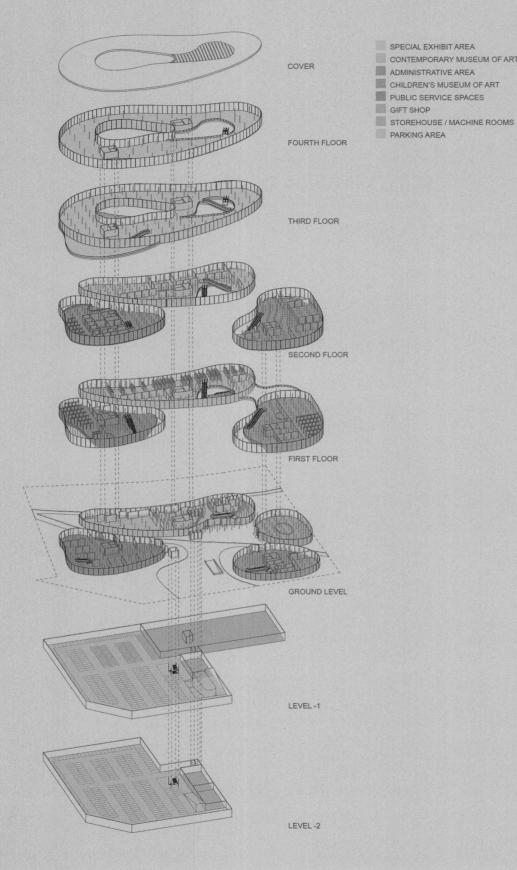

■	SPECIAL EXHIBIT AREA
■	CONTEMPORARY MUSEUM OF ART
■	ADMINISTRATIVE AREA
■	CHILDREN'S MUSEUM OF ART
■	PUBLIC SERVICE SPACES
■	GIFT SHOP
■	STOREHOUSE / MACHINE ROOMS
■	PARKING AREA

Design strategies
Put together, each space is part of a large connected organism that expands and extends itself through the site, acting as a filter and transitional space between public and art.

设计策略
每一个空间就好比一个大型的相互连接的有机体的一部分。当它们被组合在一起时，会不断的向该项目地块扩展、延伸，就像一个过滤器，是公众与艺术之间的一个过渡空间。

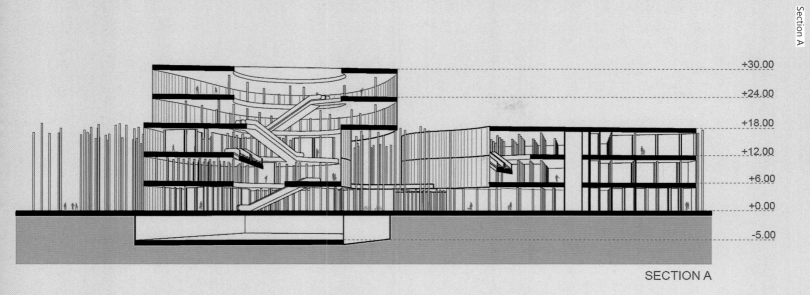

SECTION A

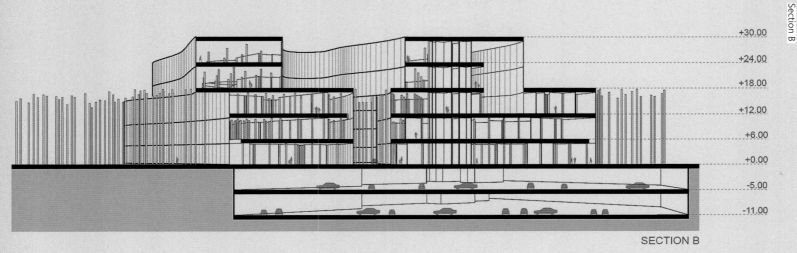

SECTION B

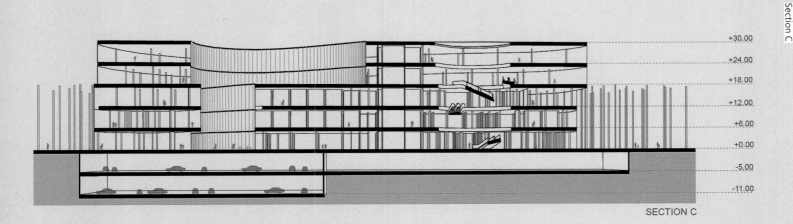

SECTION C

A forest of columns, extending from the topography of the exterior into the interior spaces, acts as the device that delineates space; each programme has a specific uniform distance between each column. This creates a unique atmosphere throughout the museum in which spaces are only roughly defined.

林林总总的立柱一直从博物馆外部空间延伸至内部空间，就像是一台机器，勾勒着空间造型。每一栋建筑物与每一根立柱之间的距离是特定而统一规定的。如此一来，整个博物馆的氛围变得独一无二，内部空间的界限十分模糊，营造出一种独特的朦胧美。

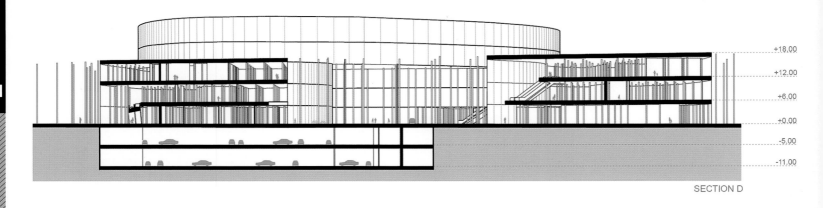

SECTION D

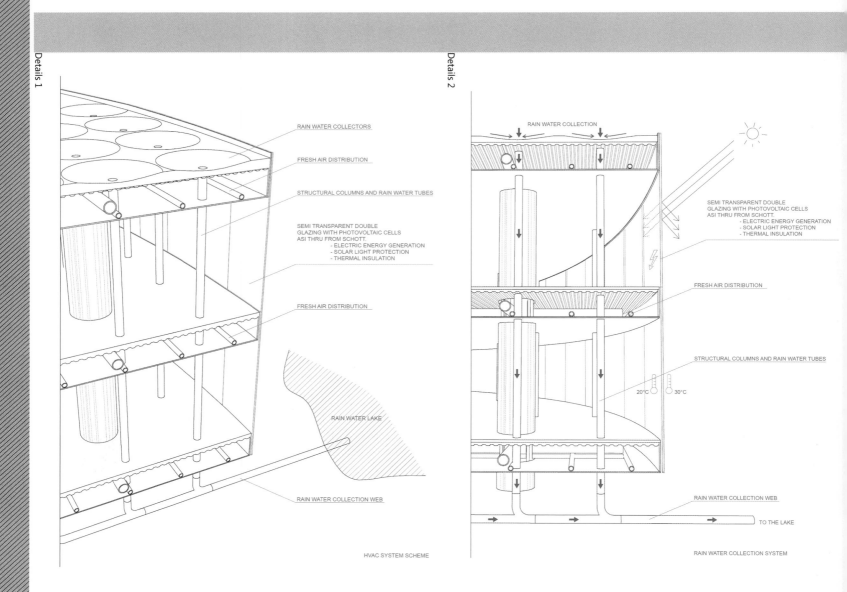

HVAC SYSTEM SCHEME

RAIN WATER COLLECTION SYSTEM

The first three levels of the museum sit in between the forest of columns around the building. As the visitor moves upwards towards the upper two floors that house the main galleries, this forest starts to disappear. Upon arrival, the visitor can enjoy a space that floats above all else.

博物馆最底部三层恰好位于围绕大楼设置的立柱森林的中间。当游客向上移动至博物馆的较高两楼层时，到达主艺术画廊的同时，立柱森林也将慢慢地从视野里消失。身处这个"漂浮"与一切之上的空间里，游客们将获得不一样的体验。

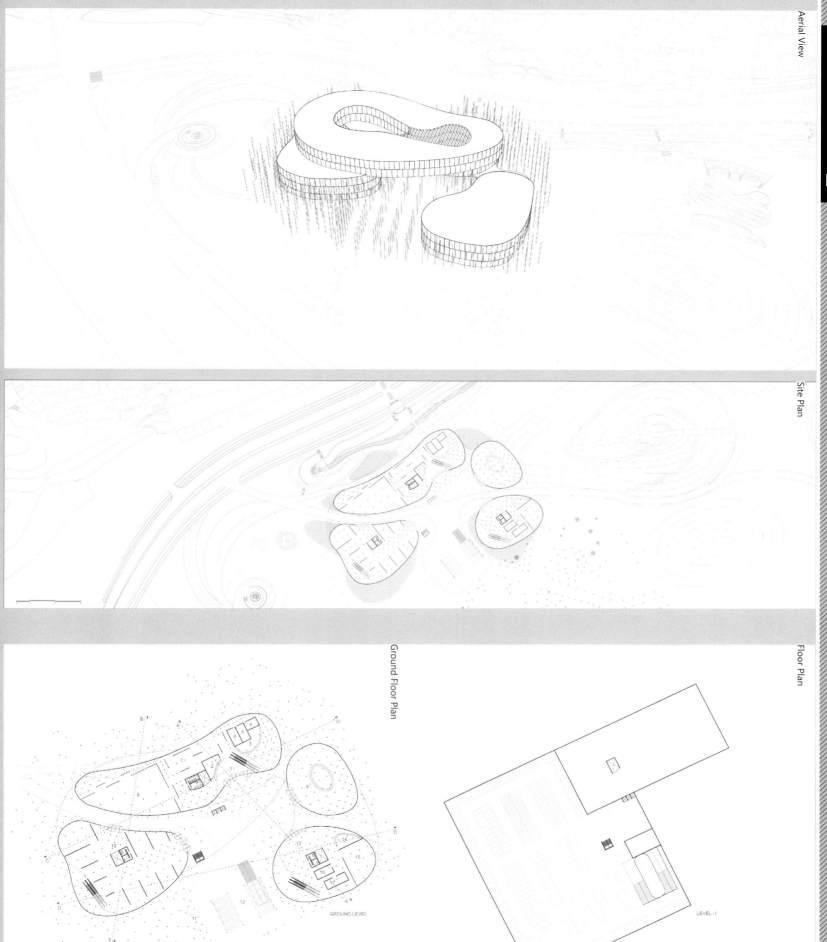

Aerial View

Site Plan

Ground Floor Plan

Floor Plan

These main gallery spaces are defined above all else by flexibility. Here, the column grid is the most open, allowing museum curators ample room to design exhibitions and sequences that can suit any particular need. It is this level of flexibility and openness that will make the New Taipei City Museum of Art a unique public forum for art, learning, and culture.

这些主画廊空间的主要特色便是其设计的灵活性。在这里，立柱骨架是最为开放式的设计，可给予博物馆管理者充足的空间来设计安排能适应任何独特需求的各种展览和顺序。正是设计的灵活度和开放度使得新台北市艺术博物馆将成为一个独特的公众艺术、学术以及文化论坛。

1st Floor Plan | 2nd Floor Plan
3rd Floor Plan | 4th Floor Plan

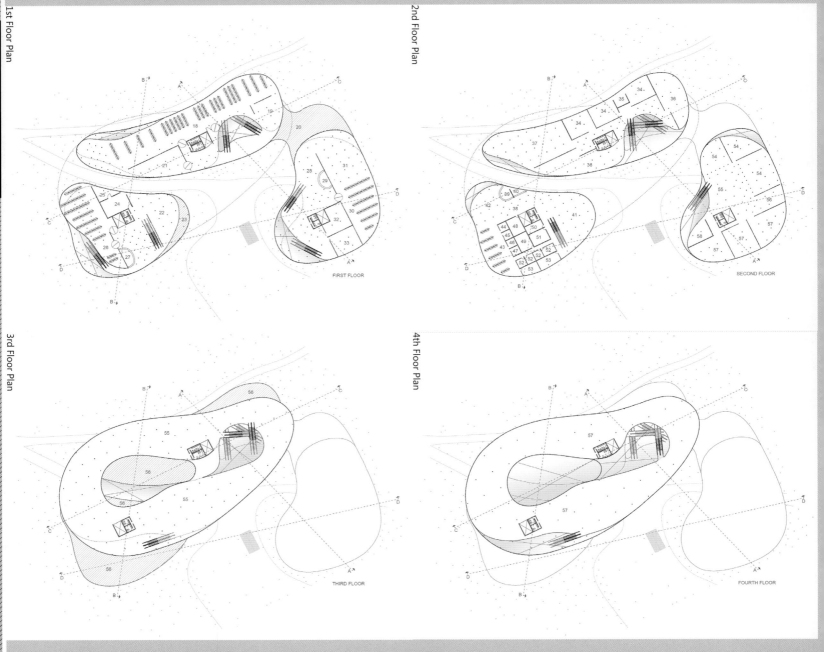

PROBLEM / 问题

① CULTURAL AND HISTORY BACKGROUND
② THE LIMITATIONS OF THE ENTITY
③ OPEN LEARNING SPACE

1. 文化和历史背景
2. 实体上的限制
3. 开放的学习空间

New School of Architecture, Royal Institute of Technology (KTH)
新建筑学院，皇家技术学院（KTH）

Tham & Videgård Arkitekter works
Tham & Videg rd Arkitekter 作品

Architect: Tham & Videgård Arkitekter
Client: Akademiska Hus.
Location: Stockholm, Sweden
Area: 9,140 m²
Function: Education

设计公司：Tham & Videgård Arkitekter
客户：Akademiska Hus.
地点：法国里昂
面积：9 140 平方米
功能：教育

DESIGN REQUIREMENTS
设 计 要 求

The narrow and triangular site in relation to a large building program was a chalenge. In addition to this the adjacent buildings are protected and a cultural heritage so the new building had to respond in a good way to its local context.

在大型建筑项目周围构建狭窄、三角形基地是一个挑战。此外，邻近建筑为受保护文化遗产，因此新建筑必须对此作出良好准备。

① 为了让新学院跟基地更融合,建筑外形圆润,包括一个下沉庭院和一个屋顶平台,将庭院打造成一个连续的空间。

② 深红色钢板外立面与现有深红色砖建筑相呼应。

③ 建筑内部设计的弧形墙面营造出一种自由流动的连续空间,增强了开放感而不是闭合性。

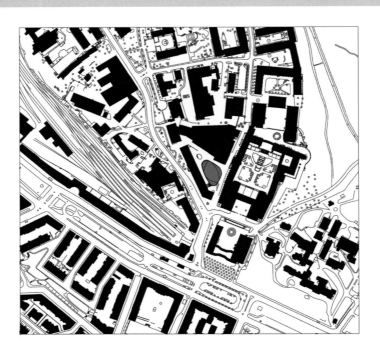

Location Plan

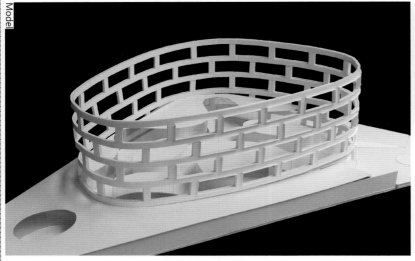

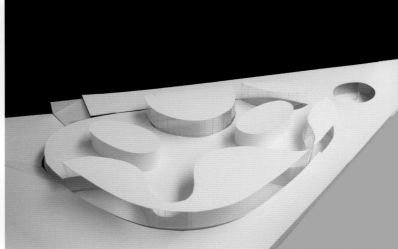

Model

Project Overview / 项目概况

The site on the KTH campus, with its very tangible cultural and historical context and its physical limitations, could be described as the opposite of a blank slate (Tabula Rasa). The new school is inserted into an existing courtyard space with existing pathways and is located adjacent to Erik Lallerstedt's original and quite monumental brick buildings from the early twentieth century.

建筑场地在皇家技术学院校园内,有着十分切实的文化和历史背景,也有实体上的限制,所以它可谓是白板(尤指洛克哲学纯洁质朴的思想状态)的反面。新的学院大楼坐落在一个已有的庭院内,道路完成,与20世纪早期Erik Lallerstedt设计的原始、相当具有纪念意义的红砖建筑相毗邻。

1. The building has a round outline, including a suken courtyard and a roof terrace to make the new college be in harmony with the site.

2. The deep red steel facade echoes the deep red brick buildings.

3. The curved walls within the building interior create a free-flowing continuous space, which enhances the openness rather than closeness.

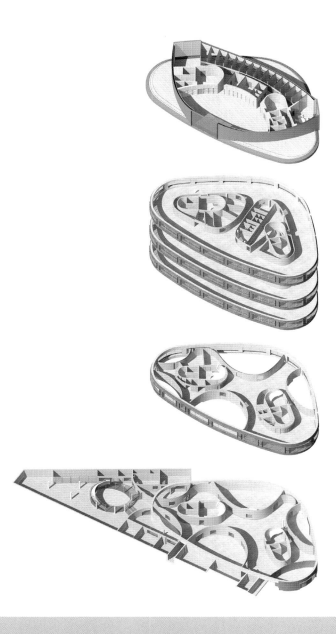

结构分析图
Structure Analysis

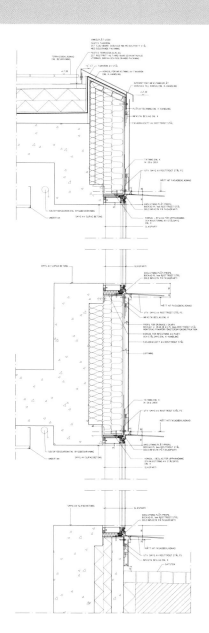

Details

By studying different footprint and volumes of the building the architects realised a curved shape making the building appear small in relation to its actual size. This observation was the starting point for defining the inner logic of the structure, organising space and programme by using convex and concave walls, creating mass and void. In addition to this the deep red Corten steel exterior related to the dark red brick of existing buildings in a good way.

通过研究建筑物的不同的足迹和体积，我们发现使用弯曲形状，可以使建筑物比实际尺寸看起来更小。这种发现是定义结构内部逻辑、使用凹凸墙面组织空间和项目、创造质量和空隙的起始点。此外，深红色钢外饰面与现有建筑的深红砖构造也相得益彰。

Based on the logic of a free campus layout that encourages movement, the idea is to accomodate and encourage circulation within the building and all around it as a way of thoroughly integrating and anchoring the new school to the site. With its rounded contours and a total of six floors, the school building includes a sunken garden and a roof terrace, while cultivating the character of the courtyard as one continuous space. The deep red Corten steel exterior relates to the dark red brick of existing buildings.

校园布局较为松散,可以鼓励运动,基于同样的逻辑,想法是鼓励学生在室内和建筑周边多多走动,以此让新的学院和基地更融洽地结合在一起。建筑外形圆润,共六层,教学楼包括一个下沉庭院和一个屋顶平台,将庭院打造成一个连续的空间。深红色科尔顿外部钢结构与现有深红色砖建筑相呼应。

内部设计稳健、灵活。弧形墙面营造出一种自由流动的连续空间,增强了开放感而不是闭合性。建筑结构内道路贯穿,随处可见室外景观,相比传统机类建筑,其空间条件要更像是一处景观。入口层有一系列的两层高空间、工作室和展区,体现主入口处空间宽敞,其高度为两层,可做开放式演讲厅使用。外形为曲折的宽敞通道,贯穿建筑。

The interior is designed to be robust and flexible. Curving walls create a free flow of contiguous space that enhances the sense of openness rather than enclosure. Views and paths are extended through the structure with spatial conditions more akin to a landscape than a traditional institutional building. At the entrance level a series of double height spaces, the atelier and exhibition area, designate a generous main entrance that also doubles as an open lecture hall. It is in the form of a broad passage meandering through the building.

A deep floor plan creates an opportunity of extensive glass use in the surfaces of the facade. It endows the building with a high degree of generality, offering lavish amounts of light and transparency, while maintaining the climate and energy efficiency of the whole building.

平面图较深，创造机会可以在外立面的表面大量使用玻璃。赋予建筑高度的通用性，提供充足的光线和透明度，同时保持整栋建筑的气候和能源效率。

PROBLEM / 问题

1) STRENGTHEN THE URBAN CONTEXT
2) CREATE NEW PUBLIC SPACE
3) EXTEND THE LANDSCAPE ONTO THE SITE

1. 加强城市文脉
2. 创造新的公共空间
3. 扩展地点所在景观

Wuzhong Museum
吴中区博物馆

Studio Link-Arc
works
Studio Link-Arc
作品

Architect: Studio Link-Arc
Client: Wuzhong District Culture and Sports Bureau, Wuzhong Urban Construction and Development Company
Location: Suzhou, China
Area: 14,461 m²
Function: Museum

设计公司：Studio Link-Arc
客户：苏州市吴中区文化体育（文物）局，苏州市吴中区城区建设发展有限公司
地点：中国苏州
面积：14 461 平方米
功能：博物馆

DESIGN REQUIREMENTS
设 计 要 求

Located within a modern development outside the historic centre of Suzhou, the Wuzhong Museum site exists at the intersection of history and modernity. Within the masterplan, the project site is located between a reconstructed canal, a new landscaped park, and a major new pedestrian mall and also enjoys a view of the ancient Baodai bridge. The project brief asked to respond to the complex site, the history of the area, and the culture of the Wu people, whose art will be on display within the museum.

吴中博物馆位于历史名城苏州城外的开发区，这里是历史与现代的交叉路口。根据总体规划，项目地点位于重建的水道、新建的景观公园和一个主要步行街的交汇处，同时也与古色的宝带桥毗邻。按照项目大纲要求，我们需要应对该地点的复杂性，反映区域历史和吴人文化，并且在该博物馆中也将展示吴人文化。

THE SOLUTION / 解决方式

① 以合理方式布局项目。本项目的大量公共项目位于底层，以便于从公园和公共街道进入博物馆。博物馆的行政办公室位于二楼，在画廊和底层储藏区之间。画廊位于上层，以便于环境光线进入和观赏景观。

② 提升画廊和办公室：我们提升项目画廊和办公室空间，以便于在建筑前两层之上构建公共空间。

③ 连接画廊，延伸景观：画廊项目由两部分构成，以方便灵活设计。较低空间边缘向下延伸，以连接到景观和公共空间。

④ 扩展公园，构建画廊：我们从项目底部楼层挖出一个较大空隙，从而形成一个连接公园的广场。画廊呈现曲线外观，以反映各种地点影响。

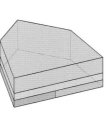

Locate the Program in a Logical Manner
均匀分布各个功能

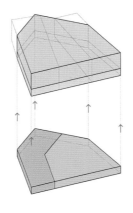

Lift the Galleries and Offices
抬起展览与办公空间

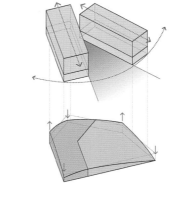

Articulate the Galleries and Extend the Landscape
打开展览空间并延伸景观

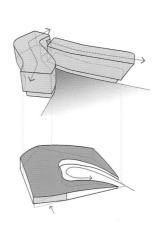

Extend the Park and Form the Galleries
调整展览空间和绿化屋面形态

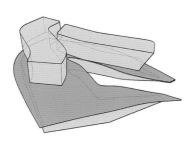

Final Form
最终形式

B1 Floor Plan / 负一层平面图

LEVEL B1 (-6.00) 地下一层

A　FACILITIES / 设备及车库
A01　Public Entrance / 公共入口
A03　Air Conditioning Room / 空调机房
A04　Fire Pump Room / 消防泵房
A06　Water Pump Room / 水泵房
A07　Boiler Room / 锅炉房
A08　Building Maintenance Suite / 楼控用房
A09　Elevator Machine Room / 电梯机房
A10　Service Corridor / 服务通道
A18　Service Elevator / 服务电梯
A19　Exit Stair / 防火楼梯
A20　Loading Office / 装卸办公室
A21　Network Center / 网络中心
　　　Storage / 储藏室

B　PUBLIC SERVICES / 公共服务区
B13　Public Elevator / 公共电梯

Project Overview / 项目概况

Situated just south of Suzhou's historic city centre, the Wuzhong Museum is the centrepiece of a redevelopment plan for the Dantai Lake Scenic Area. Located at an important juncture within the masterplan, the site is bordered by a landscaped waterfront park to the north, a reconstructed canal and commercial development to the south, and a major pedestrian plaza to the east. After understanding these disparate conditions, Studio Link-Arc began designing the Wuzhong Museum with a number of specific goals. The first goal was to strengthen the

吴中区博物馆用地遥望古城坐拥新城，是澹台湖景区城市更新计划的核心。毗邻作为世界文化遗产的宝带桥与京杭运河，是吴中地区悠长历史和多样文化的缩影。针对吴中地区的文化特性和博物馆建筑的性质，我们制定了一系列具体的设计目标。一、加强场地城市文脉，延续悠长的历史，以此诉说当下。二、丰富和拓展公共生活，聚合人气、活化场地。三、延续北侧的公园景观绿化。与此同时，我们也希望创造有强烈表现力的建筑图景，重塑城市空间。

Locate the Programme in a Logical Manner: The project's extensive public programmes are located on the ground floor for easy access from the park and public streets. The museum's administrative offices are located on the second floor, between the galleries and the ground floor storage areas. The galleries are located on the upper floors to access light and views of the context.

Lift the Galleries and Offices: The architects lifted the project's galleries and office spaces to create a new public space on top of the building's first two floors.

Articulate the Galleries and Extend the Landscape: The gallery programme is formed into two bars in order to allow flexibility in programming. The edges of the lower volume are pushed downward to connect to the landscape and public space.

Extend the Park and Form the Galleries: The architects carved out a large void from the lower floors of the project to create a new entry plaza from the park. The galleries are formed into curvilinear masses that respond to various site influences.

LEVEL 1 (+0.00) 首层

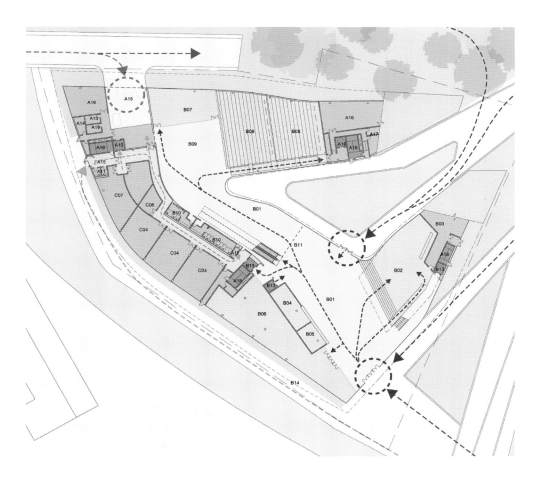

SPACE KEY | 空间排布

A   FACILITIES / 设备及车库
A11   Restroom / 卫生间
A15   Loading Dock / 装卸区

B   PUBLIC SERVICES / 公共服务区
B01   Public Concourse / 大厅
B02   Restaurant/Cafe / 餐厅
B03   Kitchen / 厨房
B04   Bag Check / 存包处
B05   Ticketing / 售票处
B06   Museum Store / 博物馆商店
B07   Seminar Room / 教学互动体验室
B08   Auditorium / 多功能报告厅
B09   Prefunction Space / 后台准备室
B10   Public Restrooms / 公共卫生间
B14   Bicycle Parking / 自行车停车

C   STORAGES / 藏品库房区
C04   Research Room / 研究室
C06   Reading Room / 鉴赏室
C07   Restoration Room / 藏品修复室

site's contemporary urban context. The second goal was to extend and promote the public space of the masterplan. The third goal was to extend the landscaped space of the park to the north. All of these aims were considered in conjunction with a larger desire to create an expressive work of architecture within the new masterplan.

XXX Floor Plan

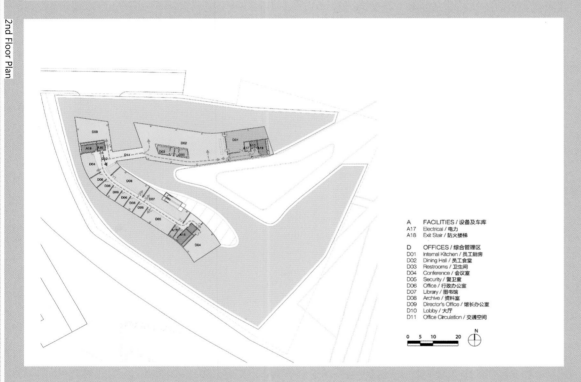

A	FACILITIES / 设备及车库	
A10	Service Elevator / 服务电梯	
A11	Restroom / 卫生间	
A12	Janitor's Room / 值班室	
A13	IT Closet / 通信设备接入机房	
A17	Electrical / 电力	
A18	Exit Stair / 防火楼梯	
A20	Network Center / 网络中心	
A21	Storage / 储藏室	
B	PUBLIC SERVICES / 公共服务区	
B12	VIP Space / 贵宾接待	
B13	Public Elevator / 公共电梯	
C	STORAGES / 藏品库房区	
C02	Basic Storage / 普通库房	
C03	Rare Storage / 精品库房	

2nd Floor Plan

A	FACILITIES / 设备及车库	
A17	Electrical / 电力	
A18	Exit Stair / 防火楼梯	
D	OFFICES / 综合管理区	
D01	Internal Kitchen / 员工厨房	
D02	Dining Hall / 员工食堂	
D03	Restrooms / 卫生间	
D04	Conference / 会议室	
D05	Security / 警卫室	
D06	Office / 行政办公室	
D07	Library / 图书馆	
D08	Archive / 资料室	
D09	Director's Office / 馆长办公室	
D10	Lobby / 大厅	
D11	Office Circulation / 交通空间	

3rd Floor Plan

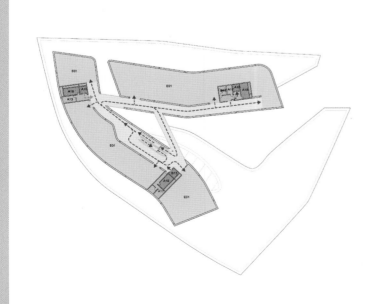

A	FACILITIES / 设备及车库	
A10	Service Elevator / 服务电梯	
A13	IT Closet / 通信设备接入机房	
A17	Electrical / 电力	
A18	Exit Stair / 防火楼梯	
B	PUBLIC SERVICES / 公共服务区	
B10	Public Restrooms / 公共卫生间	
E	EXHIBITIONS / 陈列展览区	
E01	Main Gallery / 主展厅	

The design team arrived at the form via a series of simple gestures, which, when taken together, create an iconic building which is uniquely responsive to its site. The first major gesture was to lift the landscape along the site's south and west edges. This gesture creates a strong urban response and extends the park onto the site. The museum's public components, including the lobby, restaurant, shops, and educational/seminar spaces are located beneath this landscape, expanding the public space of the masterplan onto the project site. A deep cut into this lifted landscape creates an entry courtyard for the museum and allows light to penetrate within.

设计团队通过一系列相辅相成的简洁操作，创造对场地有独特回应的标志性建筑。首先，我们将需要最佳景观的展陈功能从地面抬起，并把公园景观绿化延续进博物馆场地中，回应了城市现状。大厅、餐厅、商店和教学互动体验室等公共空间位于景观下方，是总体规划中公共空间在基地中的延伸。被抬升的景观上的深切口形成了博物馆的入口庭园，并使得光线能够到达建筑内部。

Section AA

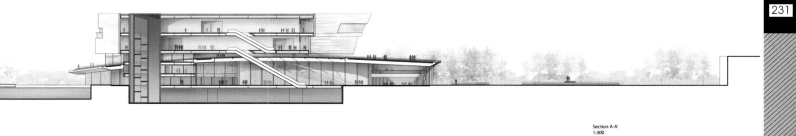

Section A-A'
1:300

Details 1

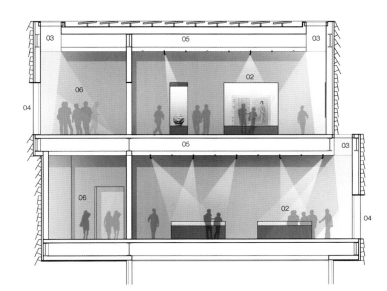

01   Gallery Entrance / 展厅入口
02   Vitrine / 展柜
03   Skylight / 天窗
04   Window / 窗
05   Drop Ceiling / 悬挂天花
06   Corridor / 走廊

Details 2

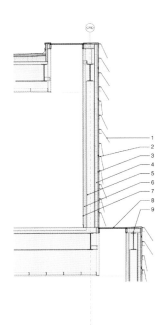

1. Natural Clear Anodized Aluminum With Bead Blast Finish
2. Continuous Horizontal Panel Clip
3. Aluminum Metal Back Panel Over Vapor Barrier And Exterior Sheathing
4. Mineral Wool Insulation
5. Structural Steel Column
6. Interior Acoustic Batt Insulation
7. Interior Gypsum Board With Plaster Finish
8. Skylight Insulated Glazing System With Ceramic Grit
9. Structural Steel Beam

1. 天然氧化铝表面喷砂处理
2. 连续水平板夹
3. 铝金属背板及防潮层和外护套
4. 矿棉保温层
5. 结构钢柱
6. 室内声学保温层
7. 石膏完成面的室内石膏板
8. 陶瓷熔块隔热玻璃天窗
9. 结构钢梁

两个东西向的线性展厅体量漂浮于抬高的景观上方。北侧展厅体量朝向宝带桥，而南侧展厅则朝向规划的新区。两个体量由封闭的玻璃廊桥相连，使游客在参观过程之中仍能欣赏周边景色。

线性的展厅使内部空间极具灵活性，提供多种展陈布置可能性。沿展厅边缘布置的天窗将漫射的北向天光带入室内。展厅的体量被包裹在闪耀的金属表皮下，在特定节点打开，使游客能够欣赏到公园周围的景色。

Above the lifted landscape, the galleries are expressed as two linear floating volumes, oriented roughly east-west. The volume to the north is gently curved to gesture towards the park, while the southern gallery volume engages the public mall to the west of the site. The volumes are connected via enclosed glass bridges, which allow visitors to engage with the site while embedded in the gallery experience.

两个东西向的的线性展厅体量漂浮于抬高的景观上方。北侧展厅体量朝向宝带桥，而南侧展厅则朝向规划的新区。两个体量由封闭的玻璃廊桥相连，使游客在参观过程之中仍能欣赏周边景色。

线性的展厅使内部空间极具灵活性，提供多种展陈布置可能性。沿展厅边缘布置的天窗将漫射的北向天光带入室内。展厅的体量被包裹在闪耀的金属表皮下，在特定节点打开，使游客能够欣赏到公园周围的景色。

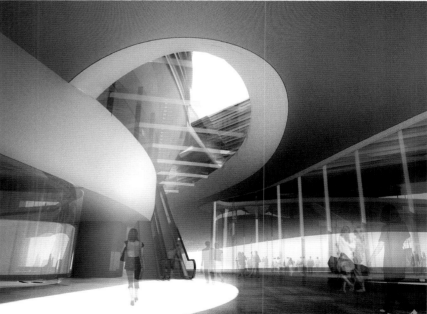

The linear nature of the gallery spaces optimises their flexibility, allowing multiple exhibition configurations. Linear skylights along the perimeter of the gallery volumes bring diffuse north light into the spaces. The gallery volumes are enclosed in a shimmering metal skin that shades the facade, but is punctured in specific locations to allow visitors to view the landscape beyond the park.

PROBLEM / 问题

① INTEGRATION WITH ENVIRONMENT
② FUNCTION
③ SUFFICIENT SUNLIGHT

1. 融入环境
2. 功能性
3. 当地充足的光照

Cité Du Corps Humain
CITé DU CORPS HUMAIN

Big
works
Big
作品

Architect: Big
Client: Ville de Montpellier
Location: Montpellier, France
Area: 7,800 m²

设计公司：Big
客户：Ville de Montpellie
地点：法国蒙皮利埃
面积：7 800 平方米

DESIGN REQUIREMENTS

设 计 要 求

The project is rooted in the Medical College of Montreal and its world-renowned medical school, exploring the human body from an artistic, scientific and societal approach through cultural activities, interactive exhibitions, performances and workshops.

① 建筑师将整个项目建筑的外观设计成一条迂回盘曲的纽带，形成一个浮沉于城市与公园之间的、无缝连接的连续统一体。

② 建筑一共被划分为八大功能区，主功能区皆沿一根主轴线分布成一个个单独的展览馆，独立又相互统一成一个整体。

③ 设计一层百叶窗系统，保护该项目建筑免受过多热辐射和紫外线的伤害。

PROGRAM

The building's program is grouped into eight major functions with the reception hall in the center.

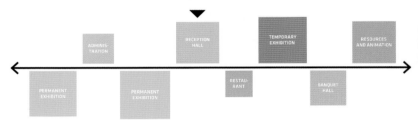

LINEAR ORGANIZATION

The functions are organized along a main axis, allowing the building to merge with its surroundings - creating views to the park, access to daylight, and optimizing internal connections.

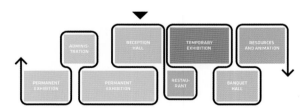

FROM LINEAR ORGANIZATION TO COMPRESSION

The organization of the functions are compressed in order to remain within the site boundaries. For practical, functional and flexibility reasons, all functions are located on one level. This compression creates connections between the functions which, if organized linearly, would not be possible.

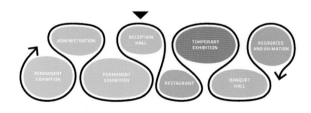

FROM COMPRESSION TO ORGANIC SHAPES

By multiplying the interfaces between the spaces, the shape becomes more functional, catering to the needs of the building - an adaption that results in a more fluid and organic shape, in osmosis with its environment.

Project Overview / 项目概况

Project overview

The Cité du Corps Humain is conceived as a confluence of the park and the city — nature and architecture–bookending the Charpak Park along with the Montpellier city hall.

项目概况

Cité du Corps Humain 项目被设计成为一个公园与城市的融合体，并真正做到自然与建筑物的完美结合。该项目建筑毗邻夏帕克公园以及蒙特利埃市政厅。

Design strategies

Like a seismic fault line, the architectural crusts of planet earth are lifted and mingled to form an underlying continuous space of caves and niches, lookouts and overhangs. A series of seemingly singular pavilions that weave together to form a unified institution – like individual fingers united together in a mutual grip.

Rather than a single perimeter delineating an interior and an exterior, the facade is conceived as a sinuous membrane meandering across the site, delineating interior spaces and exterior gardens in a seamless continuum oscillating between the city and the park. The roofscape of the Cité du Corps Humain is conceived as an ergonomical garden – a dynamic landscape of vegetal and mineral surfaces that allow the park visitors to explore and express their bodies in various ways – from contemplation to the performance, from relaxing to exercising, from the soothing to the challenging.

设计策略

整个项目建筑的外观就像一条地震断层线，在地球上通过抬升、融合的作用，最终形成一个下凹式的连续空间，似洞穴，又似壁龛，时而起伏，时而高悬。表面看去，该项目建筑是一个个单独的展览馆，实际上它们却又相互统一成一个整体，就像是交叉抱合的双手那样。

建筑师们并没有使用单一的外缘线定义建筑的内外部轮廓，而是将整个项目建筑的外观设计成一条迂回盘曲的纽带，在大地上蜿蜒飘舞，从而勾勒出内部空间和外部公园的轮廓线，形成一个浮沉于城市与公园之间的、无缝连接的连续统一体。Cité du Corps Humain 项目建筑的屋顶被设计成为一个符合人体工学原理的花园——这个动态的景观花园的表面，像植物又像矿物质，游客们可来到这个屋顶花园，以多种形式探索并表达自己的身体所想，或沉思，或表演，或放松，或锻炼，既是一种舒缓身心的方式，又是对身心的一次挑战。

striated facade with layers that bend from horizontal to vertical in a seamless transition. Like a functional ornament adapted to its native climate the facades of Cité du Corps Humain resemble the patterns you find in a human fingerprint – both unique and universal in nature.

建筑的外观设计采用了条纹元素，水平与垂直方向之间的过渡无缝自然，更具层次感。就像是一件能适应本地气候条件的功能性装饰物，Cité du Corps Humain 项目建筑的布局模式与人类指纹的分布很相似，既独特，又自然而普遍。

PROBLEM / 问题

1. **PROTECT LOCAL ENVIRONMENT**
2. **TRADITIONAL CULTURE**
3. **DIFFERENT SPACE DIMENSIONS**
4. **VARIOUS PLACES**

1. 保护当地环境
2. 传统文化
3. 不同空间尺度
4. 多重场所

Beijing River Creative Zone
北京妫河·建筑创意园区

UNStudio
works
UNStudio
作品

Architect: UNStudio
Location: Yanqing, Beijing, China
Area: 145,119 m²
Site Area: 210,152 m²
Function: Creative zone

设计公司：UNStudio
地点：中国北京市延庆县
建筑面积：145 119 平方米
占地面积：210 152 平方米
功能：创意园区

DESIGN REQUIREMENTS
设计要求

The Beijing Creative Zone represents the urban typology needed for a new open learning environment in "Urbanity in Nature".

北京创意园区代表了一个"于自然中的都市"的新型开放式学习环境所需的都市形态。

① 景观浓缩在充分利用场地自然条件的同时，也使得绿化带面积达到最大。

② 建筑公园中，画家工作室和亭阁以中国传统园林"一步一景"的方式进行合理组织。

③ 创意园区的设计包含两个层面：绿化带和都市肌理。边缘的都市肌理主要由两条不同但紧密联系的带状地区交织而成。

④ 园区将成为实验项目和新型设计产品的国际橱窗。通过实施产品生命周期中的四个环节——"思考、发展、生产与销售"形成连续完整的产品销售步骤。

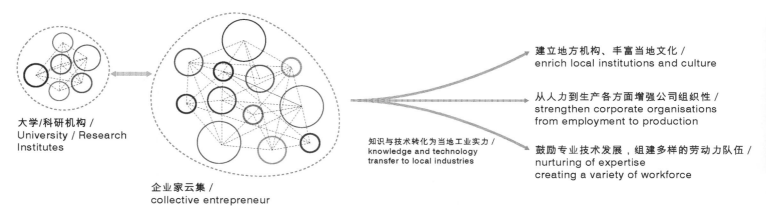

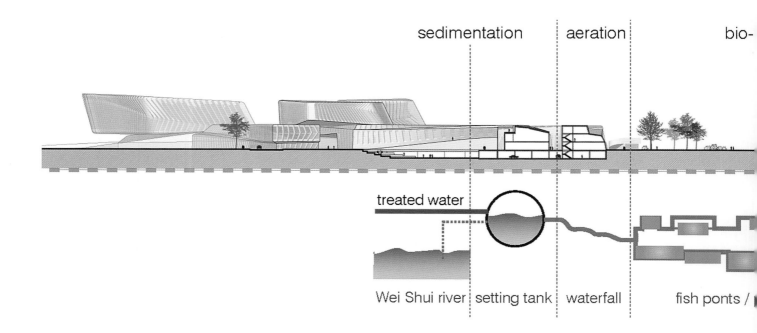

Project Overview / 项目概况

Design concept

The theme of Urbanity in Nature refers to the preservation of the natural qualities of the Yanqing site, located 200km from Beijing, whilst accommodating a certain essential level of urban life density. The design for the Beijing Creative Zone encompasses compression on a dual scale; the greenbelt and the urban fabric.

The compression of the landscape maximises the greenbelt whilst simultaneously capitalising on the natural quality of the site.	**1**
In the creation of an Architectural Park, ateliers and pavilions are organised on the basis of the Chinese garden's emphasis of "One View Per Step".	**2**
The design for the Creative Zone encompasses compression on a dual scale; the greenbelt and the urban fabric. The urban fabric of the edge is formed by the interweaving of two distinct yet highly interconnected strips.	**3**
The parkland becomes an international showcase of experimental projects and emerging design practices. By implementing the four clusters of a products life cycle "Think, Develop, Produce and Sell" as a continuous loop of activity, one creates a platform for all steps of the process towards product sales.	**4**

Rise and fall of water level

water level +476m

water level +477.5m

water level +479m

water level +477.5m

water level +476m

Water Levels Analysis

natural filtration | UV desinfection | reuse

reclaimed water

wetlands | recreation waterbody | Weu Sui river

设计理念

"自然中的都市风格"这一主题旨在保护延庆县（距离北京市 200 千米）当地的自然特征的同时，容纳一定密度的都市生活风格。北京妫河·建筑创意园区的设计将把绿带与城市肌理两个尺度结合起来。

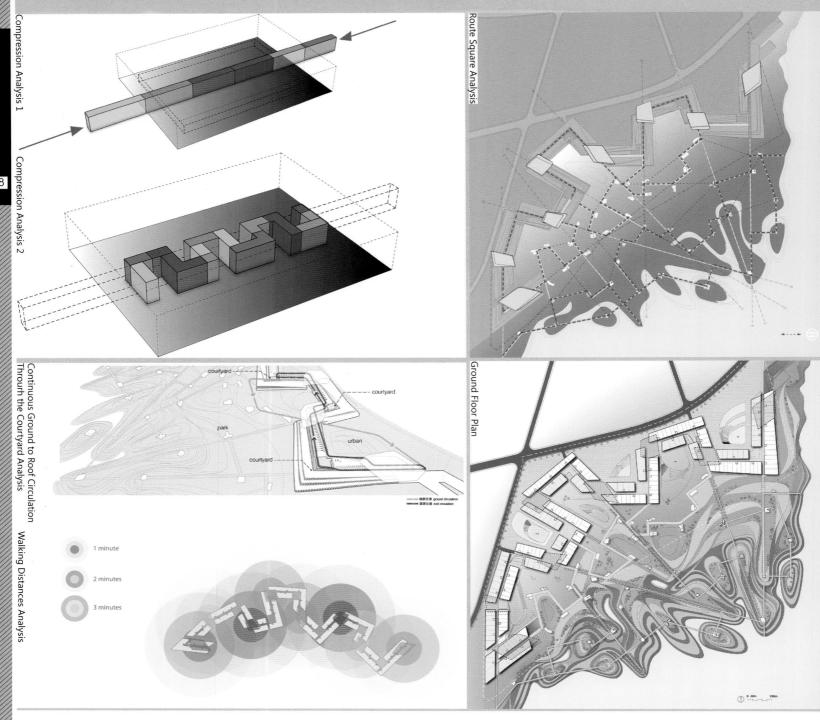

Design strategies
The compression of the urban fabric creates a tightly woven and intensified urban environment. This generates compactness and programmatic friction as well as a clear border between city and nature; the urban edge. The urban fabric of the edge is formed by the interweaving of two distinct yet highly interconnected strips - the Creative Industry Based strip and the Supporting Facilities strip, and consequently creates a series of open spaces ranging from internal streets to courtyards and plaza entrances to public squares. This arrangement provides changing degrees of privacy and interaction, thereby encouraging spontaneous learning.

设计策略
通过对城市肌理的浓缩，创造出一个高度编织和紧密的城市环境。如此一来，在城市与自然之间既形成清晰的界限又具有紧凑性——城市边缘地带。城市边缘地带的城市组织实际上是由两个不同但又相互联系的"带"组成的——即创意

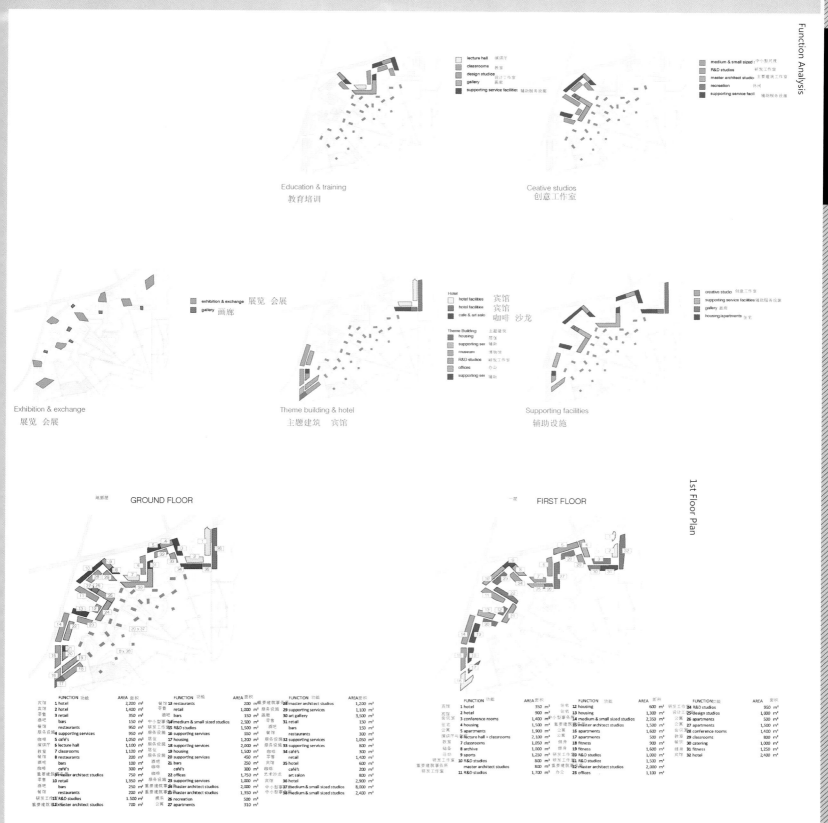

The compression of the landscape maximises the greenbelt whilst simultaneously capitalising on the natural quality of the site. The increased greenbelt and parkland provide an expanse of green environment where local and foreign creative professionals can work, organise and participate in events, as well as indulge in leisure activities in the scattered ateliers and pavilions.

园区带和基础设施带。同时还营造了从街道到庭院再到公共广场的一系列连续的开放空间。这种安排设计方式为人们不同程度的隐私和互动需求提供了空间，因此也可以鼓励人们的自发学习。
浓缩景观在充分利用本地自然特色的同时，最大限度地扩展了绿带，增加的绿带和公园为本地和外来的创意人士工作、组织和参加活动以及画室和亭、阁的娱乐活动提供了绿色环境。

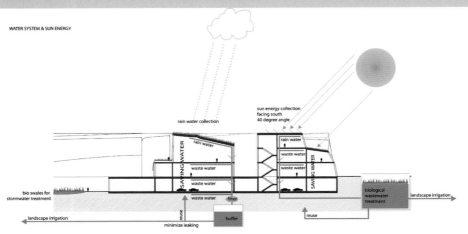

Sustainability watersystem & Sun energy

In the creation of the Architectural Park, these ateliers and pavilions are organised on the basis of the Chinese garden's emphasis of "One View Per Step". The parkland becomes an international showcase of experimental projects and emerging design practices. The ateliers and pavilions sit in the landscape as sculptures in a garden, improving legibility, navigation and orientation, as well as securing the existence of medium and small sized companies. The ateliers of emerging designers would be used as galleries to present their work as well as featuring their design philosophies and ideas on formal expressions, sustainability or material as a built artifact.

The building of pavilions, as nodes in the park, as rest stops and as outdoor event spaces, allows for a wider range of designers to participate in the urban plan through the design of small, experimental projects. The lifespan of the ateliers and pavilions varies from semi-permanent to temporary respectively. The park as a showcase will therefore be continuously refreshed.

在建筑公园中，画家工作室和亭、阁皆以中国传统园林"一步一景"的方式组织。公园变成试验项目和即将面世的设计实践的国际展示场。画家工作室和亭、阁就像是坐落在景观中的雕塑一般，提升了易辨性和方向感，同时也为中小规模公司提供了安全的工作场所。画室可以作为新兴设计师们展示作品、表达设计思想（对形式表达，可持续性发展或者手工艺品材料的观点）的场所。
亭、阁作为公园中的小品，可以用作休息站和室外活动空间，允许设计者通过一些小而试验性的设计加入到城市设计之中。画室和亭、阁可以是临时的，也可以是半永久的，因而公园将成为一个不断处于更新变化之中的天然展示场。
北京创意园区的设计参考了伦敦经济学院的开放式校园模式。交织在城市肌理中的校园允许公众自由地进入，进行学习和活动交流——这是开放式校园设计的一个典型范例。通过分析此校园的模式，我们了解到主要建筑物（比如说学术报告厅和图书馆）以及用作知识交流的互连公共空间的重要性。尽管在创意园区的设计过程中，强调的是教育交流，但是主要目的仍然是在城市背景下通过主要设置一些集会和交流场所来营造一种开放的学习环境。庭院穿插在以内部街道形式呈现的户外集会空间里——在这里，可以举办一些小型活动、论坛和展览等。这样一来，庭院就将成为进入城市、公园广场和内部街道的主要联系空间，同时也将成为人与人之间相互联络的场所。
创意园区的全方位产品循环周期将成为项目可行性的主要支柱。通过实施产品生命周期循环中的四个环节——"思考、发展、生产与销售"，并将其作为活动的连续回路，那么我们将创造出一个完成这一过程所有步骤的平台，并最终迈向产品销售。这样一个可行的循环链将为市场进行比较灵活的转型提供有利条件，而且还能将知识创造与创意的发展以及实际产品的开发联系起来，并最终使产品的展览与销售成为可能。
这些变化的元素构筑了城市与公园相接触的平台，基于绿带的城市边缘上相互混合的地带展现了城市类型学，而这正是自然之中的城市所必须具有的新式开放型学习环境。

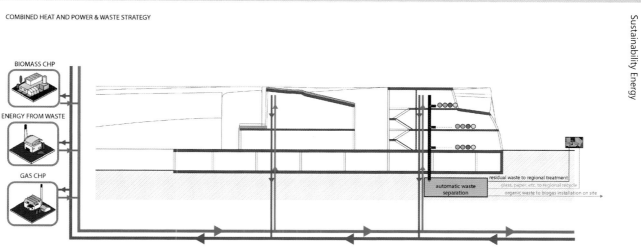

COMBINED HEAT AND POWER & WASTE STRATEGY

Sustainability Energy

The London School of Economics campus is seen to have potential as a reference for the design of the creative zone. It is an example of an open campus that is interwoven in the urban fabric and thus allows for the public to freely interact with the many learning facilities and events. Analysis of the campus model shows the importance of key buildings (such as a lecture hall and a library) and interlinked public areas for the exchange of knowledge. Although emphasis is made on educational exchange in the design for the Creative Zone, the aim is to create an open learning environment within an urban context through the placement of key meeting and exchange locations. The courtyards punctuated along the internal street form outdoor gathering spaces where small events, forums and exhibitions can take place. These courtyards will be the major junctions between access to the urban and park squares and to the internal street thereby creating an interface between the occupants and the public.

The all-encompassing production cycle of the Creative Zone will be the backbone of the feasibility of the project. By implementing the four clusters of a products life cycle "Think, Develop, Produce and Sell" as a continuous loop of activity, one creates a platform for all steps of the process towards product sales. The feasibility of such a chain allows for flexibility in market transformation as well as linking the creation of knowledge with the development of an idea, actual product development and finally exhibiting and selling the piece.

These various components, consisting of urban and park facing platforms with the interwoven programmatic mixed strips lined on the urban edge by the greenbelt represent the urban typology needed for a new open learning environment in Urbanity in Nature.

PROBLEM / 问题

1. **HOT, RAINY AND DAMP CLIMATE**
2. **NORTH-SOUTH LONG AND NARROW LAND BLOCK**
3. **STRONG SUNLIGHT IN WEST-EAST DIRECTION**
4. **VARIOUS FUNCTIONS WITH LESS CONNECTION**
5. **TYPICAL TRADITIONAL SURROUNDING ARCHITECTURE, DIFFICULT TO FIT THE PUBLIC BUILDING**

1. 炎热多雨潮湿气候
2. 地段狭长型南北走向
3. 东西晒严重
4. 功能类型多而相关性弱
5. 周边建筑传统风格强烈，且难以适合公共大型建筑尺度

Xinglong Visitor Centre
兴隆访客中心

Atelier Alter
works
时境建筑
作品

Architect: Atelier Alter
Local Team: China Northeast Municipal Engineering Design & Research Institute
Client: Xinglong Overseas Chinese Farm
Location: Xinglong, Hainan Province, China
Area: 15,781.7 m²

设计公司：时境建筑
本地团队：中国市政工程东北设计研究总院
客户：兴隆华侨农场
地点：中国海南兴隆
面积：15 781.7 平方米
功能：

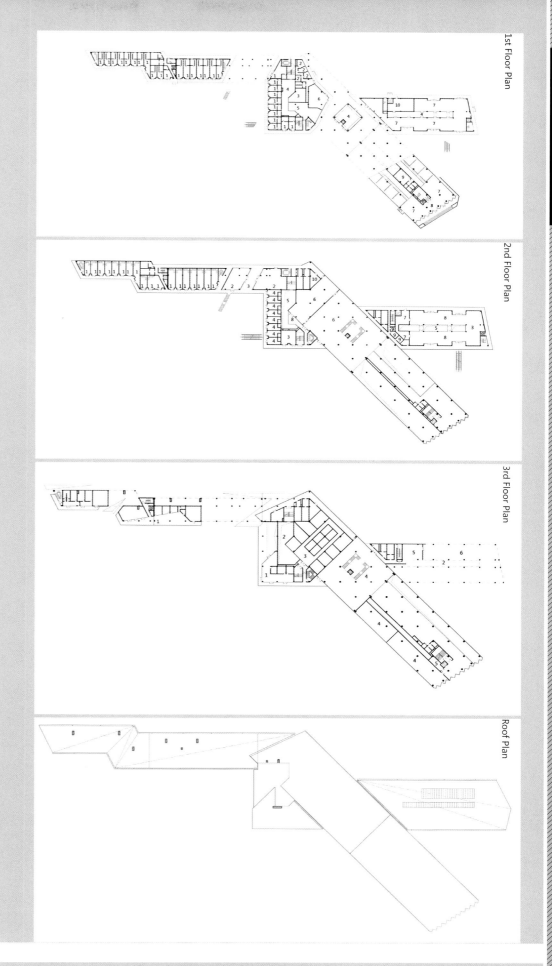

characteristic history and culture legacies and delicious food. From the very start, architects are actually focused on this design from the perspective of a city.

沿海修养名镇的历史、面貌、饮食等，访客中心的建筑从一开始就是从城市角度入手的。

Section 1-1

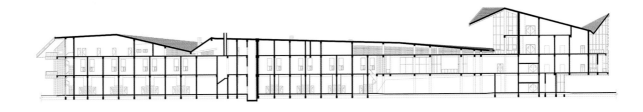

Section 2-2

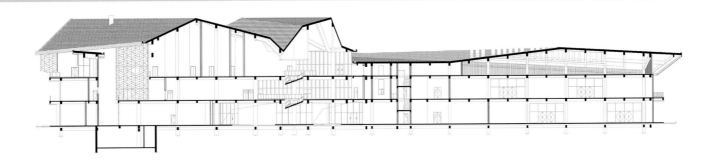

Section 3-3

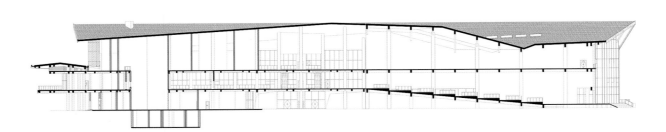

Section 01

Section 02

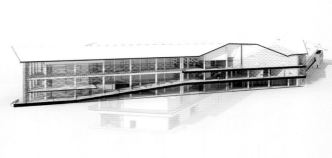

Design strategies

Architects proposed a linear zigzag building that places most of the building volume in the middle and leaves the two ends with landscape and lighter constructions. Meanwhile, building volume is pushed back from the street to create a linear plaza joint by triangular semi-enclosed open spaces. Approaching from both ends, the architecture opens up to the street in a dramatic way.

Elevation

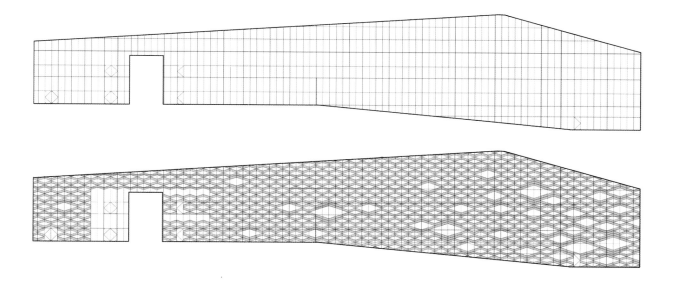

Details

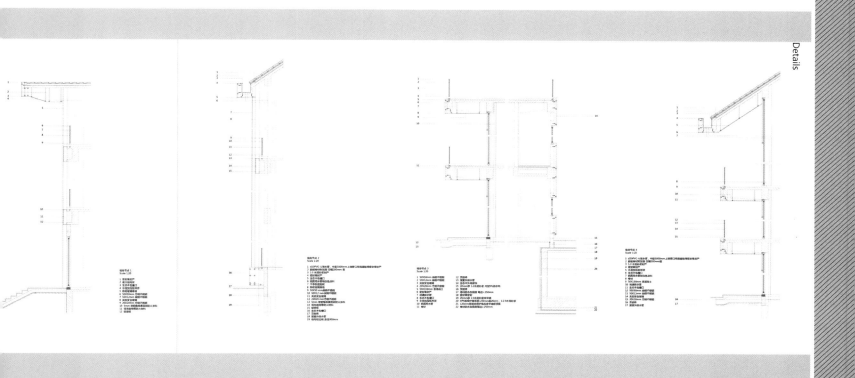

设计策略

建筑师设计了一个线性锯齿状建筑,将主体集中在地段的中部,在地段两端设置了较为轻的体量,以容纳两端的景色;同时两侧的临街面,建筑体量退了一定距离,形成了一连串的三角形半开放广场。从不同的方向接近这个建筑都有截然不同印象。

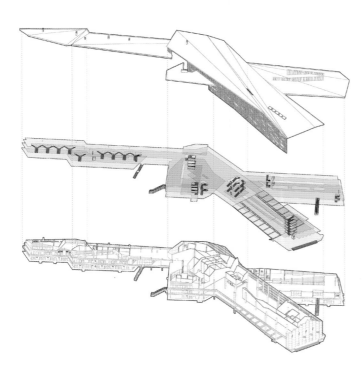

The functionality of the visitor centre includes exhibition, administration management, shops and caterings. To respond to the mixed-use programme, rather than giving a definitive formal statement, architects try to recreate the impression about the fishing island and make it into stunning spaces through juxtaposing different types of fishing lodges from the island on top of each other in various scales and directions. The project retains the memory of the fishing island in the making of architecture. Architects have transformed their respect towards traditional weaving craftsmanship into designing a double curtain wall system to shelter from strong sunlight and radiation from both east side and west side; slightly-rolling sun shields are helpful to timely ventilate the heated air with excellent heat-dissipation performance. However, the structure expression is also analogous to the tropical plants on the island, highlighting slender tree trunks and spreading crown over the head.

访客中心的功能包含展览、旅游区的行政管理、商店和餐饮，建筑师没有设定一个明确的形式的宣言，而是针对复杂多样的功能对渔岛的传统印象进行了再创造，通过对小屋原型在不同方向、不同尺度的叠加、切割，形成了惊人多样的空间。本项目在建筑的工艺方面也保留了对渔岛的回忆，通过对传统编织手工的敬意，建筑师将其转译成双层幕墙系统以遮挡来自东西两面的阳光和辐射；略微起伏的遮阳板有助于热空气的及时流动，提供更好的遮阳板散热；结构形态蕴含了本地热带植物的意向：纤细的树干和在高高的头顶展开的树冠。

PROBLEM / 问题

1. **NEW ECONOMIC PHENOMENON IN TAIWAN**
2. **TRADITIONAL AND CONTEMPORARY ARCHITECTURE**
3. **LOCAL ECOLOGY**
4. **NEW LIFESTYLE**

1. 台湾新经济形势
2. 传统与当代建筑
3. 当地生态
4. 新兴生活方式

Hsinchu Stone Village
竹岩溪村

Gianni Botsford Architects and Mole Architects
works

Gianni Botsford Architects and Mole Architects
作品

Architect: Gianni Botsford Architects, MOLE Architects
Location: Hsinchu, Taiwan, China
Area: 45 hectares (according to TLDC's description)
Function: Resort hotel, serviced apartments, spa, chapel and luxury villas

设计公司：Gianni Botsford Architects, MOLE Architects
地点：中国台湾新竹
面积：45 公顷
功能：度假酒店，服务式公寓，水疗中心，教堂和豪华别墅

DESIGN REQUIREMENTS

设 计 要 求

The original design requirement was for a retreat style hotel and serviced apartments set in the countryside. The architects developed this into the concept of a hill village that contains many elements that contribute to an interesting and varied daily life for the village including a hotel, serviced apartments, villas, clubhouse, cafés and restaurants, a wedding chapel and a varied landscape.

原希望能在乡间创建一个静养型酒店和公寓。后来，我们将这一设计理念转化为山间乡村设计。在村庄内融入诸多元素：一家酒店、公寓、别墅、俱乐部、咖啡厅、餐馆、举报婚礼的小教堂以及其他景观，为乡间生活增色不少。

THE SOLUTION / 解决方式

① 除台湾领先科技院校和高新科技研究机构外，新竹早已开始探索新门路如保健和医疗旅游等，来维持繁荣经济的可持续性。新竹石头村内设酒店、公寓、私人别墅以及各种购物、饮食、运动和娱乐设施，是对台湾这一新经济形势做出的创意响应。

② 设计团队对客户要求做出的回答是：在充分尊重传统台湾历史和当地客家建筑风格的前提下，开发一个具有当代特色的山间小镇。如此庞大的开发项目需要在几层独立并偶有联系的庭院间分别完成，建筑所用材料既有新型材料，也有传统材料，包括砖头、木材和石材等。

③ 规划小村坐落在由新竹平原突起的山麓丘陵上南向斜坡上。建筑与景观应该完全融入现有环境，而不是摧毁现有景观，推出平坦的平台。此外，项目还应在秉承传统中国设计原则的同时，也考虑西方国家建筑习惯，积极应对各种生态挑战和机遇。

④ 设计大胆设想了新型生活方式，建筑中的现代公寓和高级度假酒店创造出一个独具特色的"乡村"。处处是景观怡人的乡间小路，还配有一个中央广场。商业区位于私人别墅旁，站在"云"咖啡馆与餐馆前，即可俯瞰整个中央广场。举办婚礼的小教堂就设在花园内，一条条小径通往茶园迷宫。

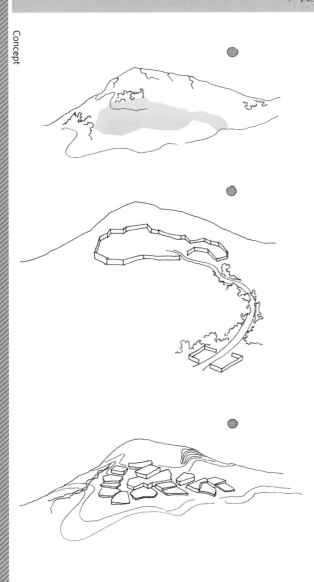

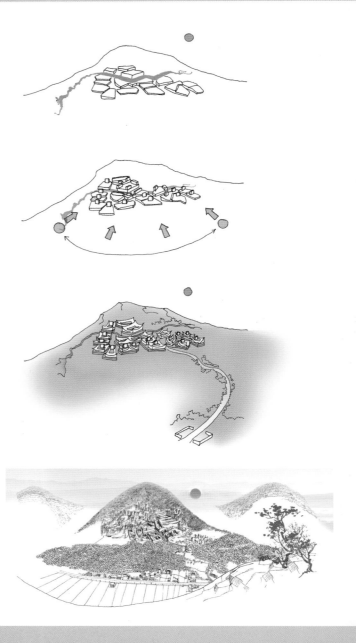

Project Overview / 项目概况

Upper level planning

Hsinchu Stone Village is a creative response to this new economic phenomenon, and the brainchild of the Taiwan Land Development Corporation (TLDC) – a Taipei-based company which promotes and engages in rural redevelopment and urban renewal. The project is among the TLDC's first forays in large-scale town and country planning. Eager to create an

上层规划

竹岩溪村，是台湾土地开发公司 (TLDC) 所独创的创新设计规划，用来迎接这个新兴的经济现象的智慧结晶。位于台北的 TLDC 长期致力促进乡村发展及都市更新。此规划案是 TLDC 中许多突破以往的大尺度的城乡规划之一。企图创造一个创新暨国际级的开发案，将整合旅馆、饭店式管理公寓、私人别墅以及多元性的购物、美食、运动及休闲娱乐的设施。TLDC 从海外寻求协助，尤其是英国的专业设计团队。

In addition to the Taiwan's leading science and technology universities and high-tech research institutes, Hsinchu has begun to explore new means of sustaining its booming economy such as health and medical tourism. Hsinchu Stone Village is a creative response to this new economic phenomenon comprising hotels, serviced apartments, private villas, as well as numerous facilities for shopping, eating, exercising, and relaxation.

The design team's response to the client's brief has been to create a new, contemporary hill town with vernacular and historical references to traditional Taiwanese and local Hakka architecture. The bulk of the development shall be deployed in a series of independent and sometimes interconnected multi-storey courtyards, and built using a range of new and traditional building materials, including brick, timber and stone.

The site of the proposed village straddles a broad south-facing slope on the foothills rising from the Hsinchu Plain. Instead of destroying the existing landscape to make flat platforms, buildings and landscape shall be fully integrated, and informed by traditional Chinese design principles imbued with Western idiosyncrasy ("Chinois-Anglos-Chinois") and responsive to ecological constraints and opportunities.

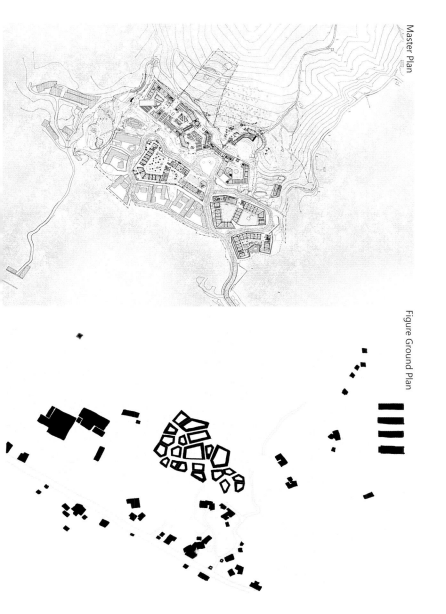

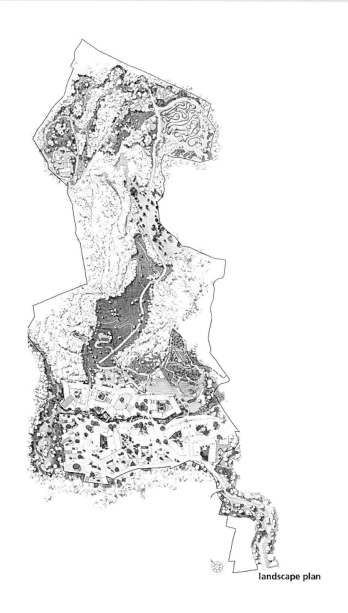

landscape plan

innovative and world-class development comprising hotels, serviced apartments, private villas, as well as numerous facilities for shopping, eating, exercising, and relaxation, the TLDC looked abroad, and to British-based designers in particular to assist them.

Structure Analysis

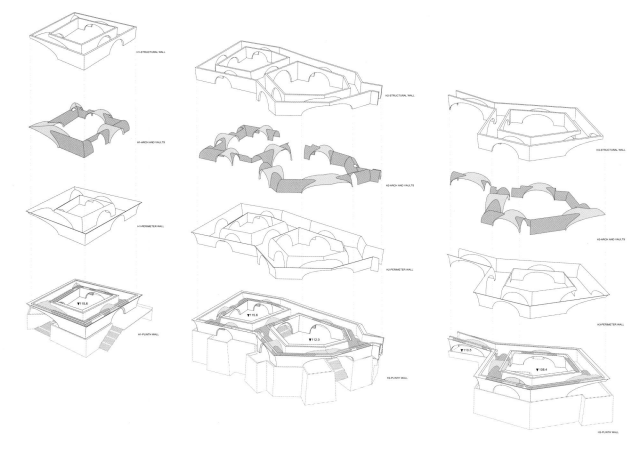

Roof Plan

Hotel Arch Reflected Ceiling Plan

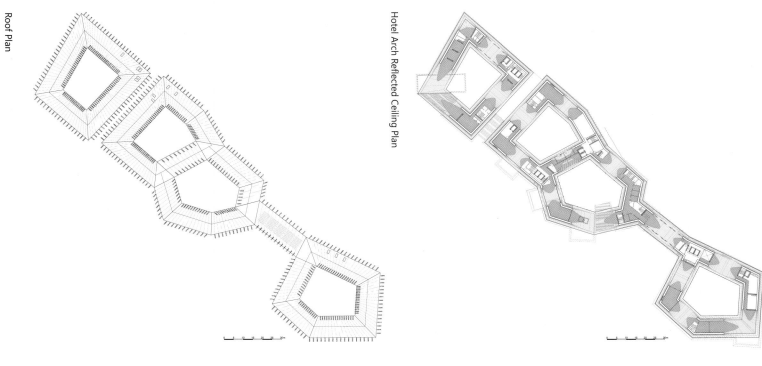

Location analysis
The site of the proposed village straddles a broad south-facing slope on the foothills rising from the Hsinchu Plain, the east and west ends of which tumble into precipitous wooded valleys. Half the 200-hectare site is flat, open and gently undulating, and half is steeply sloping and smothered with a dense mat of lush tropical vegetation. Much of the it was until recently farmed for centuries, and its upper reaches contain many relics including betel palms groves, crumbling stone terraces and orphaned camellias from abandoned tea plantations.

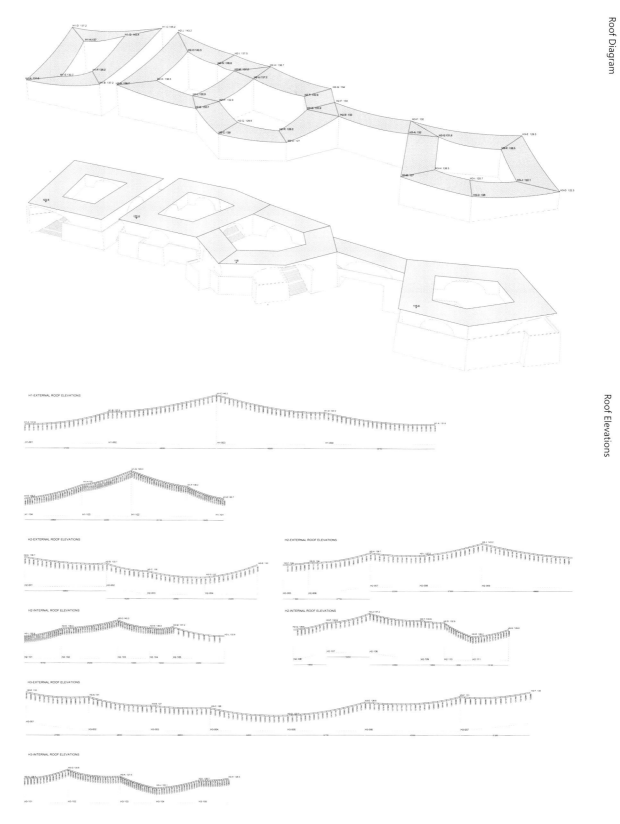

Roof Diagram

Roof Elevations

区位分析

此提案的基地座落在新竹平原中隆起丘陵宽阔的南向坡上，东西两端的地势没入陡峭茂密的山谷中。两百公顷的基地，一半是平坦宽阔且缓缓起伏；另一半是被稠密的土壤和葱郁茂盛的热带植被所覆盖的陡峭坡地。这些植被大多是几世纪以来的耕作，向上坡延伸的植被中含括被耕作过的痕迹，例如：槟榔林、外露的石台以及生长在遗弃茶园裡的山茶花。

Design strategies

The design team's response to TLDC's brief has been to create a new hill town with vernacular and historical references to traditional Taiwanese and local Hakka architecture. The bulk of the development shall be deployed in a series of independent and sometimes interconnected multi-storey courtyards, and built using a range of new and traditional building materials, including brick, timber and stone.

设计策略

此设计团队依循台湾当地传统客家建筑，创作出具有当地历史风格建筑语汇的新式山城，借此回应 TLDC 的方针。整个开发案的核心须采用一系列的现代建材和传统的砖、木、石材，创造出一连串独立且又间歇相连的合院。
新村的配置也是同样重要：建筑与地景必须完全融为一体，并且将西方特殊的建构手法（"Chinois-Anglos-Chinois"）完全展现在传统中国的设计原则中，同时以此回应环境生态限制及可行性。

The setting of the new village is no less important: the buildings and landscape shall be fully integrated, and informed by traditional Chinese design principles imbued with Western idiosyncrasy ("Chinois Anglois Chinois") and responsive to ecological constraints and opportunities.

The design imagines a new form of living where modern serviced apartments and a luxury resort hotel create a unique single "village", with landscaped paths and a central piazza. Business retreats sit alongside private villas, a "cloud" café and restaurant overlook the piazza, a wedding chapel is located in a garden, where paths lead into a tea garden maze.

此设计的想像是一种新型态的生活方式。在这里，现代化饭店式管理公寓和华丽的度假酒店藉由众多的地景小径和中央广场，造就了一个独特单一的"村落"。商务会馆紧邻在私人别墅旁、面对广场的云顶咖啡厅和餐厅、花园裡的婚礼教堂，引领进入茶园迷宫。

PROBLEM / 问题

1. **CREATE A "GATEWAY" VISIBLE FROM MAINLAND CHINA TO KINMEN ISLAND**
2. **FOSTER A NEW ERA OF FRIENDSHIP AND TRADE BETWEEN AND Taiwan, China**
3. **CONNECT TO LOCAL CULTURE AND OFFER VALUE TO THE RESIDENTS OF KINMEN ISLAND**
4. **CREATE A WORLD-CLASS PORT TERMINAL IN TERMS OF FUNCTION AND TECHNOLOGY**

1. 创建一个从中国大陆到金门岛的可见"通道"
2. 促进中国和台湾之间构建新型友谊和贸易关系
3. 联接本地文化,并未金门岛住民创造价值
4. 打造功能和技术领先的世界级港口码头

Kinmen Passenger Service Centre
金门客运服务中心

Tom Wiscombe Architecture
works
Tom Wiscombe Architecture
作品

Architect: Tom Wiscombe Architecture
Client: Kinmen County Government, Taiwan, China
Location: Kinmen Island, Taiwan, China
Area: 42,000 m²

设计公司:涌现组
客户:中国台湾省金门县政府
地点:中国台湾省金门岛
面积:42 000 平方米

DESIGN REQUIREMENTS
设计要求

The Kinmen Passenger Service centre is one of several new port terminals to be built in Taiwan, both increasing its capacity for trade and tourism, but also signalling to the world Taiwan's interest in supporting contemporary architecture. This terminal is both a gateway and destination, so it must relate to the rich local culture of Kinmen Island and tourists going there, as well as move people to other major cities in China and Southeast Asia.

THE SOLUTION
解决方式

① 该项目将港口码头转换为带有多变内饰和屋顶观景台的公共场所。

② 该建筑可以为多个分散个体或单个新个体,并且这些个体的组合,应构建一种"通道"的轮廓。

③ 大型的内部物体,可在码头内构建一个内部空间,同时也将空间进行分割。

④ 金门当地住民和游客可在该建筑的屋顶花园中参加和享受文化盛宴,且该屋顶花园连接到地面,并位于港口码头内部。

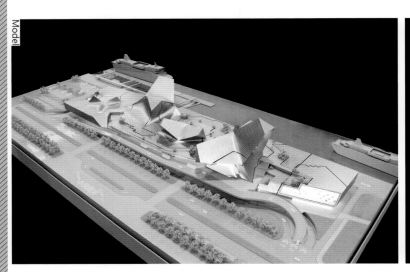

Model

Model

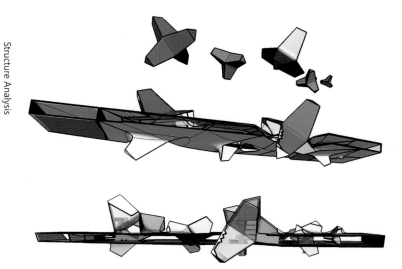

Structure Analysis

Project Overview / 项目概况

A new era of communication

This proposal for the Kinmen Passenger Service centre is based on the idea that what we are building is not just a piece of infrastructure, but also a cultural intervention. To do this, the unique history of Kinmen must be considered. Kinmenese culture has evolved from its Fujian Province traditions and formal establishment as "Kinmen" (meaning: "Golden Gate") in the 14th century, but also by becoming a multi-cultural crossing-point for trade, having been occupied by the Japanese from 1937-1945 in World War II, and then becoming part of

金门客运服务中心提案的基础是,我们建造的不仅仅是基础设施,更是一种文化形象。为了做到这一点,就必须考虑到金门的独特历史。金门文化发源于福建省传统,在14世纪时正式取名为"金门"(寓意:黄金之门),但也因其成为一个多文化交叉贸易点,在第二次世界大战期间(1937-1945)曾被日本占领,然后在1949年国共内战之后受到蒋介石国民政府的控制。从1949至1992年间,金门从一个安静的小岛逐渐演变成军事前线,小岛的生活方式彻底改变。地

① The project transforms the typology of a port terminal to become a public space with a varied interior and a roof landscape.

② The Building can be read simultaneously as several discrete objects or as one new object, and the aggregation of objects creates the "gateway" silhouette.

③ The large interior objects create an engaging interior space in the terminal, but also functionally divide the spaces.

④ Local Kinmen residents and tourists can meet and enjoy cultural festivals on the rooftop garden of the building, which is connected to the ground and the inside of the port terminal.

交織的建築圖騰 INTERWOVEN ARCHITECTURAL PATTERNS

燕尾磚斜紋
Fujian brick pattern

迷宮式斜紋
Maze pattern

交錯圖騰
Interwoven patterns

橫紋板材
Cross-grain panels

自由曲線接縫
Freeform seams

迷宮式斜紋
Diagonal grids and mazes

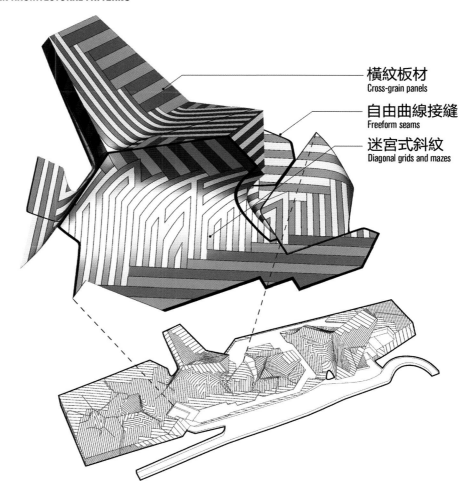

the territory of the Chiang Kai-shek's Nationalist government after the Chinese Civil War in 1949. During the period between 1949 and 1992, Kinmen was transformed from a quiet island culture into a military front line, which radically changed its way of life. Its ground became three-dimensionalised into a network of underground bomb shelters and sea tunnels for protecting people and ships, while its surface became hardened by military installations. Communication in this era of tense Taiwan-Mainland relations consisted of physical shelling, and visible propaganda slogans and lines of tanks along opposing shorelines.

底是保护人类和船舰的三维防空洞及海底隧道，地表则遍布着各种军事设施。那个时代海峡两岸关系紧张，冰冷的大炮、清晰的宣传标语和一排排坦克充当着交流的语言。

遮雨棚系統 RAIN SCREEN SYSTEM

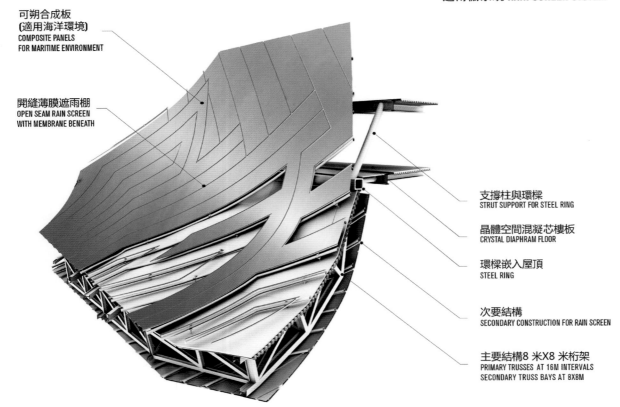

- 可朔合成板 (適用海洋環境) COMPOSITE PANELS FOR MARITIME ENVIRONMENT
- 開縫薄膜遮雨棚 OPEN SEAM RAIN SCREEN WITH MEMBRANE BENEATH
- 支撐柱與環樑 STRUT SUPPORT FOR STEEL RING
- 晶體空間混凝芯樓板 CRYSTAL DIAPHRAM FLOOR
- 環樑嵌入屋頂 STEEL RING
- 次要結構 SECONDARY CONSTRUCTION FOR RAIN SCREEN
- 主要結構 8 米 X 8 米桁架 PRIMARY TRUSSES AT 16M INTERVALS SECONDARY TRUSS BAYS AT 8X8M

屋頂綠化步道 GREEN ROOF PROMENADE

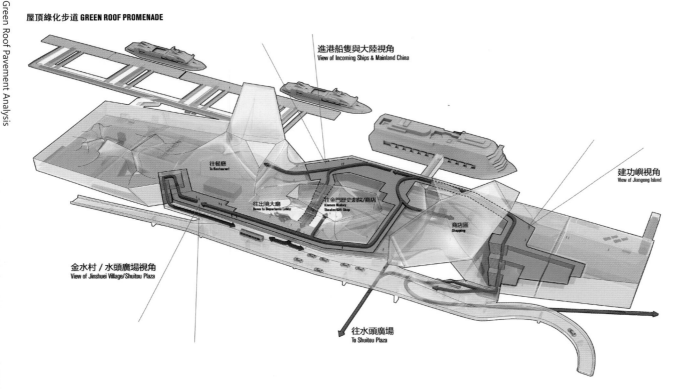

- 進港船隻與大陸視角 View of Incoming Ships & Mainland China
- 往餐廳 To Restaurant
- 往出境大廳 Down to Departures Lobby
- 往金門歷史戲院/商店 Kinmen History Theater/Gift Shop
- 商店區 Shopping
- 建功嶼視角 View of Jiangong Island
- 金水村 / 水頭廣場視角 View of Jinshuei Village/Shuitou Plaza
- 往水頭廣場 To Shuitou Plaza

Nevertheless, there is another more nuanced story here, that of deep understanding between the Kinmenese and the Mainland Chinese, a sense of shared cultural and economic history and destiny, a sharing of resources (such as water during wartime, and now, in the form of a planned pipeline), and a shared appreciation of the historical significance of the Island and its sublime natural habitats and architecture.

然而, 此处还有述说着一个更加微妙的故事, 有关于金门华人和大陆华人之间的深刻理解, 共同的文化经济历史和命运、共享的资源（如战争时期的水资源及现在计划修建的供水管道）、对金门岛屿历史地位、自然环境和建筑风格的认同感。

现在, 两岸关系逐渐缓和, 战争时代已经结束。需要一种全新的交流形式再次激发、挖掘金门岛的活力。文化遗产公园、野生动植物、古城和军事遗产将吸引新生代的参观者和移民。2001 年开放的中国一金门自由行是关键性的时刻,

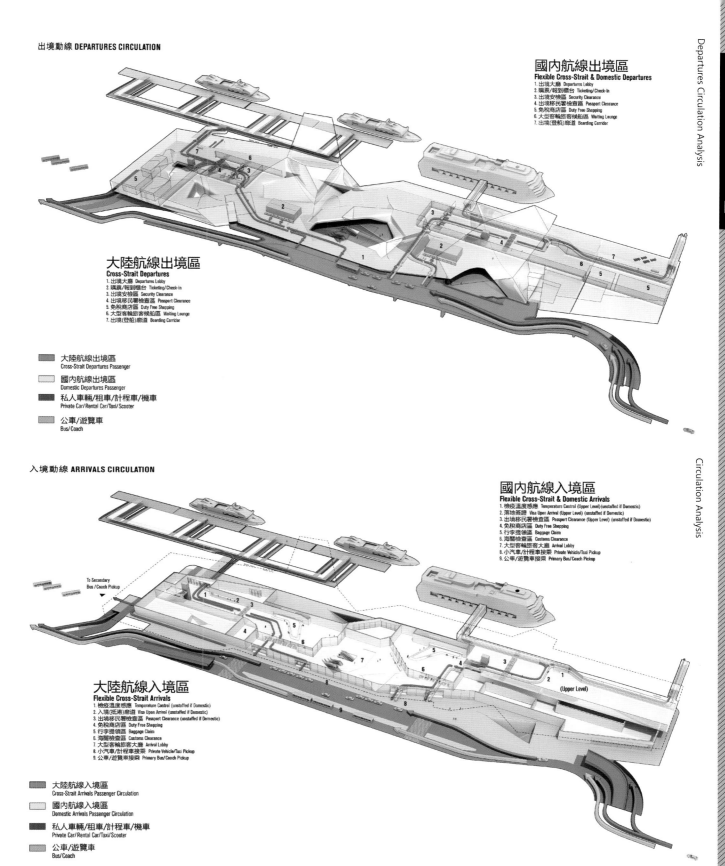

出境動線 DEPARTURES CIRCULATION

國內航線出境區
Flexible Cross-Strait & Domestic Departures
1. 出境大廳 Departures Lobby
2. 購票/報到櫃台 Ticketing/Check-In
3. 出境安檢區 Security Clearance
4. 出境移民署檢查區 Passport Clearance
5. 免稅商店區 Duty Free Shopping
6. 大型客輪旅客候船區 Waiting Lounge
7. 出境(登船)廊道 Boarding Corridor

大陸航線出境區
Cross-Strait Departures
1. 出境大廳 Departures Lobby
2. 購票/報到櫃台 Ticketing/Check-in
3. 出境安檢區 Security Clearance
4. 出境移民署檢查區 Passport Clearance
5. 免稅商店區 Duty Free Shopping
6. 大型客輪旅客候船區 Waiting Lounge
7. 出境(登船)廊道 Boarding Corridor

- 大陸航線出境區 Cross-Strait Departures Passenger
- 國內航線出境區 Domestic Departures Passenger
- 私人車輛/租車/計程車/機車 Private Car/Rental Car/Taxi/Scooter
- 公車/遊覽車 Bus/Coach

入境動線 ARRIVALS CIRCULATION

國內航線入境區
Flexible Cross-Strait & Domestic Arrivals
1. 檢疫溫度感應 Temperature Control (Upper Level) (unstaffed if Domestic)
2. 落地簽證 Visa Upon Arrival (Upper Level) (unstaffed if Domestic)
3. 出境移民署檢查區 Passport Clearance (Upper Level) (unstaffed if Domestic)
4. 免稅商店區 Duty Free Shopping
5. 行李提領區 Baggage Claim
6. 海關檢查區 Customs Clearance
7. 大型客輪旅客大廳 Arrival Lobby
8. 小汽車/計程車接駁 Private Vehicle/Taxi Pickup
9. 公車/遊覽車接駁 Primary Bus/Coach Pickup

大陸航線入境區
Flexible Cross-Strait Arrivals
1. 檢疫溫度感應 Temperature Control (unstaffed if Domestic)
2. 入境(抵港)廊道 Visa Upon Arrival (unstaffed if Domestic)
3. 出境移民署檢查區 Passport Clearance (unstaffed if Domestic)
4. 免稅商店區 Duty Free Shopping
5. 行李提領區 Baggage Claim
6. 海關檢查區 Customs Clearance
7. 大型客輪旅客大廳 Arrival Lobby
8. 小汽車/計程車接駁 Private Vehicle/Taxi Pickup
9. 公車/遊覽車接駁 Primary Bus/Coach Pickup

- 大陸航線入境區 Cross-Strait Arrivals Passenger Circulation
- 國內航線入境區 Domestic Arrivals Passenger Circulation
- 私人車輛/租車/計程車/機車 Private Car/Rental Car/Taxi/Scooter
- 公車/遊覽車 Bus/Coach

Now, as cross-strait tensions subside, and this war-time era comes to a close, Kinmen Island can be re-vitalised and re-discovered through new modes of communication. Its heritage parks, wildlife, historical villages and also its military heritage will be a draw for a new generation of visitors and immigrants. The opening of free travel between China and Kinmen in 2001 was a critical moment, now expanded by the "Three-Links" pilot programme. Recently, unprecedented high-level talks between Taiwanese and Chinese government officials have underscored the commitment of both to open communication, trade, and transportation. In this context, designing a Port Terminal for Kinmen is not only a unique opportunity but a great responsibility. The Kinmen Passenger Service centre should set the tone for the island both in terms of reflecting its complex identity and affiliations, but also in terms of presenting a vision of its future.

现在得到"三通"试点计划的进一步推动。近来，台湾政府和中国政府频繁的高层对话对开放交流、贸易和交通做出了巨大贡献。在此种环境下，金门港口码头的设计不仅是一次绝佳的机会更是一项崇高的使命。金门乘客服务中心反映金门复杂的历史身份的同时更应着眼于未来的发展方向。

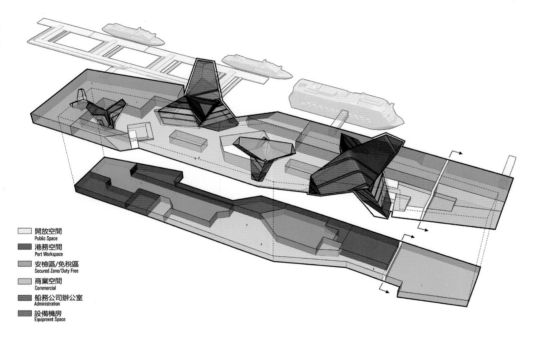

Roof silhouettes, interweaving patterns, and color

The design is intended to both symbolise a new era of open communication with Mainland China, and reflect the unique local culture of Kinmen. The strong silhouette of the design, designed to be seen from Xiamen, is characterised by dynamic figures arising from the terminal roof. The building will be a beacon for openness and transformation, literally the "golden gate" of Kinmen Island. The roof silhouette is not a generalised gesture, but rather one which is associated with the specific traditions of Kinmenese architecture. This history is defined by its complex and varying roof ridges, exemplified in the Swallowtail and Saddleback roof silhouettes seen throughout the island, originating in Taiwan and China's Fujian Province.

The tradition of complex interwoven materials and patterns in Kinmen architecture also resonates in this project. The envelope design is characterised by three interfering but complimentary patterns – freeform seams, maze-like projections, and cross-grain panels. The simultaneity of these patterns produce a heterogeneous overall effect reminiscent of local Kinmenese brickwork with its distinctive diagonal striping, as well as other unconventional juxtapositions of material scales and orientations seen everywhere on the Island. The project is conceived of as a contemporary interpretation of that craft and sensibility.

The rich palette of colours found on Kinmen Island – in its architecture, decorative arts, and in its diverse species of birds and flowers in its National Parks – are part of what makes it unique. In the design, the architects celebrate the colours of Kinmen in the interior of the Terminal. Citrus colours such as reds, oranges, yellows, and greens are used for freestanding furniture, wall graphics, roof planting, and lighting effects, to create a cornucopia of colour effects. These colours glow from inside the terminal out across the water at night. The terminal appears not as a piece of infrastructure but rather as a mirror of the interior of the Island.

顶部轮廓、交错式结构、颜色

我们的设计意在既标志着与大陆中国开放交流的新纪元，又反映了金门独特的地方文化。从厦门远眺，硬朗的轮廓清晰可见，其最大的特征是从顶端上升的动态形象。正如金门岛名字"金门"的"黄金之门"的寓意，这座大楼将成为金门开放和转型的指向标。楼顶构造不是一种通用的形式，而是取自金门建筑的特有传统。金门建筑屋脊的复杂多变述说着自己的历史，发源于台湾和中国福建省的燕尾状和鞍形屋顶在这座岛上随处可见。

善用复杂编织材料的传统和金门的建筑风格在我们这个项目中也有体现。外观设计由三个相互交错的图腾形成一横纹板材、自由曲线接缝和迷宫式斜纹。这些图腾的结合产生出一种交错的和谐感，让人联想到金门当地建筑的特色斜纹纹理，以及岛屿上到处可见的材料尺寸和方向的不规则拼合。我们的设计是对建筑工艺和敏感性的现代化诠释。

岛屿上的建筑、装饰艺术和国家公园内种类繁多的鸟类和鲜花展现出来的丰富色彩，如一个天然调色板，将金门打扮得独一无二。我们在建筑内部采用了这些金门色彩元素。独立设备、墙壁图案、楼顶喷涂和灯光效果采用红、橘、黄、绿四种颜色，创造一种丰富的色彩感。夜晚，色彩从建筑内部延伸到水面，此时，港口码头看起来更像是透明的镜子或岛屿中心，而不再是普通的基础设施工程。

Port logistics, security, and flexibility

The project is based on state-of-the-art Port Terminal design. It is focused on fluidity of passenger movement, clear security zones and passenger processing, and flexibility for future contingencies and growth. The separation between Departures and Arrivals onto two levels, with independent drop-off and pick-up zones is the most significant part of this strategy. It alleviates circulation bottlenecks both for passengers and vehicular traffic, creating uninterrupted flows, and capacity for future increases in traffic loads. Within each level, secured areas for X-ray checks, passport and ticketing checks, and waiting are clearly defined with flexible opaque and glass partitions, which can be moved to accommodate future re-organisations of the Terminal to accommodate different vessel sizes and changing proportions of domestic and international traffic.

港口物流、安全性及灵活性

本项目基于"体现艺术之美"的港口码头设计下进行。旅客的流畅通行、明确的安全区域、旅客手续办理、未来应急事件和建筑升级的灵活性都是我们关注的焦点。为了实现以上目标，我们特意将出境大厅和入境大厅分开，设计在不同楼层，

Nested crystals and sectional spaces

The building design is based on nesting five dynamically oriented Crystals into an elevated horizontal box. These Crystals push out into the box, as if stretching it, creating serene formal transitions from horizontal to vertical and hard to soft. In this way, the building appears simultaneously as a series of objects and a new whole object. On the interior, vaulted sectional spaces are created between the Crystals and the loose-fit outer shell of the building. The Departures Hall becomes a sequence of compressed and expanded spaces, creating discrete and memorable spatial experiences for travellers. The vaulted interstitial spaces also create a stack effect, where warmer air rises, drawing in cooler air from openings below. This constitutes a sustainable natural ventilation system which will be able to supplement the conventional air-conditioning system and reduce overall energy use from fall through spring.

The Crystals are arranged in plan such that they naturally organize flows of people from to ships, providing them with an automatic sense of orientation. Spatially, this is more akin to the varied sequences of spaces found in nature rather than the gridded space of the city or a Modernist free plan. This creates a calm interior experience for travellers, free of excessive passageways and signage.

The Crystals contain various programmes which need enclosure such as commercial space and restaurants on the Departures Level as well as upper level administration functions. Passengers who are shopping or relaxing on upper mezzanines of the Crystals are offered views down into the vast, bustling terminal, creating a three-dimensional spatial experience. The upper levels of the two biggest Crystals are reserved for Port Operations, Administration, and the Vessel Traffic Control centre. A double-skin envelope system is used in parts of the Crystals, with simple curtain wall construction on the inside and perforated metal panel on the outside. This allows for views out while also filtering daylight and maintaining the strong silhouette of the building. In addition, because the Crystals are buildings-within-a-building, they can contain microclimates with more tightly controlled temperature and humidity levels than in the naturally ventilated Departures Hall.

镶嵌水晶体和区域空间

五个方向各异的水晶体嵌套在水平的盒状建筑之中。这些水晶体向外伸展，实现了整个建筑从水平到垂直、外部轮廓从硬朗到柔软的完美过渡。以这种方式，同时展现出一系列建筑元素和全新组合整体的魅力。建筑内部，水晶体和宽大的建筑外壳之间形成了拱形区域空间。出境大厅是一段段压缩的延展空间，为旅客带来难忘的不连贯空间体验。拱形空间还产生烟囱效应，向上排除热空气，向下引入外部冷空气，形成可持续的自然通风系统，起到辅助空调系统，降低秋季至春季用电量的作用。

水晶体的设计将人流自然导向轮船，给他们一种下意识的方向感。从空间上来说，这更近大自然中的多样化空间，而非城市或现代规划中的网格空间。这就为旅客创造了一种平静的内部体验，没有了繁杂的通道和引导标识。

水晶内设不同的封闭空间，如商业区、出境层的酒店以及位于上层的行政区。在水晶体上层购物或休息的旅客看到下方宏伟而忙碌的港口码头，感受到天然的三维空间效果。最大的两个水晶体上层用作港口运营区、行政区和船舶交通管理中心。部分水晶体采用双层外壳，内层是简单的幕墙结构，外层是穿孔金属面板。这样在观赏外部景色的同时可遮挡强烈的阳光，还可保持建筑轮廓的硬朗风格。此外，由于水晶体是楼中楼的结构，与自然通风的出境大厅相比它更有利于控制温度和湿度，保持宜人的微气候。

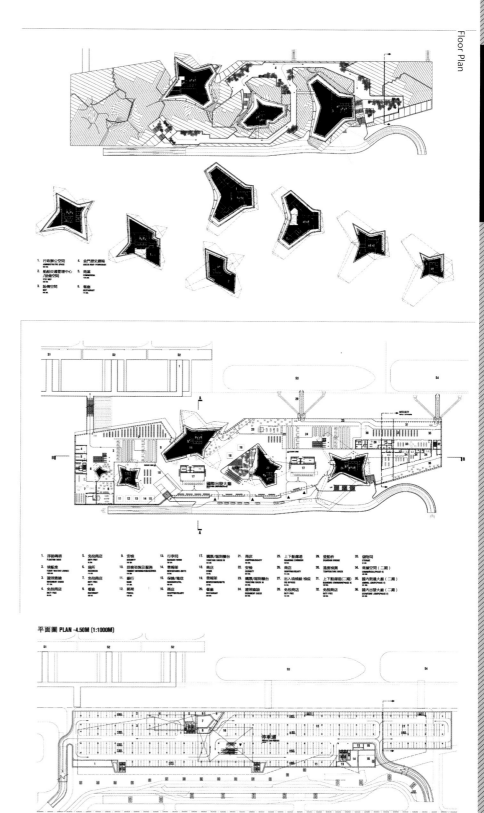

In advance of any future reconfigurations, the Terminal as designed can already accommodate both international and domestic vessels at every gate, with passport control and customs control zones in place at all points of embarking and disembarking, which, in the case of domestic travel, can simply be by-passed. While the competition brief currently only calls for international ship and ferry routes, the architects believe that the Terminal must be flexible enough to be able to respond to the fast-changing Taiwan-Mainland relations and the potentials of expanding the Kinmen Port Terminal network to include more distant ports of call in Southeast Asia and beyond.

并分别设计了独立的下客区和上客区。这样可以缓解人流和车流的交通拥堵状况，人流和车流不间断，并为未来更大的交通负荷预留空间。每一楼层的X光安检区、护照和检票口、等候区都用可移动的不透明玻璃屏风隔离开来，可移动，以适应港口未来因船舶大小和国内外交通比例的改变调整格局。

港口码头的各个泊位均已能满足国内外船舶的停靠需求，以防未来港口重构，所有登船口和登陆口设护照监管和海关监管区，国内旅客可直接绕行。而现在只需要考虑国际船舶和渡轮的航线。我们相信港口的灵活程度一定能满足正在快速变化的两岸关系以及金门港口网络辐射至东南亚甚至更远地区的潜在需求。

雨水回收系統及自然採光 RAINWATER RECYCLING AND DAYLIGHTING

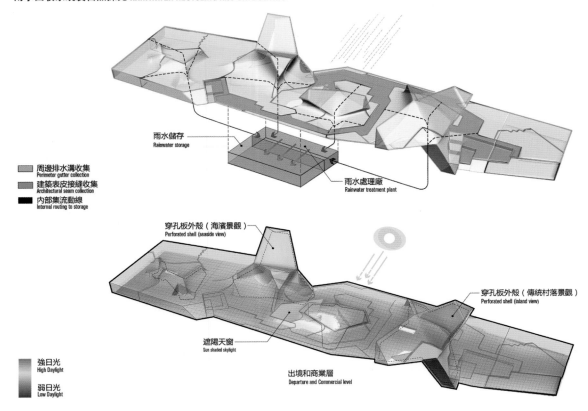

綠色能源概念 GREEN ENERGY CONCEPT

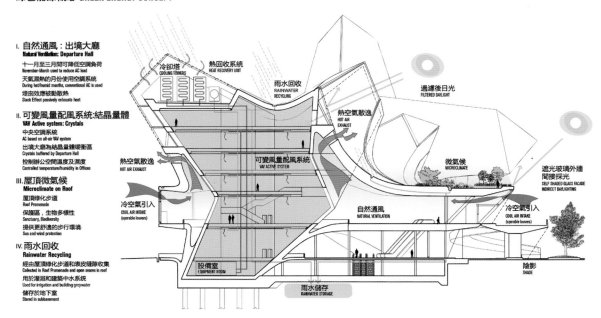

Structure and envelope

The structure of the building will be based on an economical 8X8m base column grid and a combination of reinforced concrete and steel superstructure. The proposed concrete basement is an economical rectangular shape with the base column grid, to be built using diaphragm perimeter walls to de-water the site. The foundation consists of a pile-supported mat foundation, with an internal sump system to control water penetration into the basement. The Arrivals and Departures Level floors are poured-in-place flat plate and slab-and-beam construction, and the four primary vertical cores are to be reinforced concrete. On the Departures Level, the floor is shaped in section to cantilever and tapered to the perimeter.

The Crystals are steel frame with composite deck construction for interior floors. The Crystals, in combination with their internal concrete cores, constitute the lateral bracing system for the project. The roof drapes over the Crystals, spanning from stiff rings embedded internally in the roof cavity to columns along the perimeter of the Departures Level. The roof is based on a two-way truss system with 8x8m bays, including transverse beams at 4m intervals for cladding support.

结构和外观

建筑结构基于一个经济性高的8*8米地基柱网上，结合钢筋混凝土和钢上部结构。混凝土地下室提案为矩形，有基底柱网，将使用隔膜围墙防水。地基由一个桩基底板基础组成，自带内部排水坑系统以防止水渗透至地下室内。抵达层和出境层使用现场浇灌的平板、楼板和梁构造，四个主要垂直核心为钢筋混凝土。出境层，地面部分悬臂，往周边处逐渐变窄。

The building envelope is a combination of opaque and perforated composite panel, suitable for the corrosive marine environment. The majority of the envelope is single skin, but areas which require daylight or views, especially on the upper levels of the two large Crystals, have a double-skin construction of opaque and low-e glazed curtain wall construction with perforated panel in front on a light steel construction, with access space between. The geometry of the roof is primarily planar with limited zones of one-way curvature, which can be achieved simply with edge returns to maintain panel shape. Panel sizes can be large as they are lightweight and can be delivered by barge and erected directly by barge crane. The glazed facades of the Departures Hall will be high-performance low-e glazing, optimised to insulate as well as allow the penetration of daylight.

水晶体内部楼板为钢框架复合甲板构造。水晶体与内部混凝土核心结合起来，构成了项目的侧向支撑系统。屋顶覆盖于水晶体之上，从镶嵌于屋顶空间内的硬环跨越至出境层周边的立柱。屋顶基础为一个8*8米跨度的双向构架系统，包括支撑围护结构的间隔为4米的横梁。

建筑外观结合不透明和穿孔复合板，适合这种具腐蚀性的海边环境。外观大部分为单层，但需要日照或景观，尤其是2个大水晶体的上层部分，则为双层不透明、低辐射幕墙构造，在轻钢构造前放置穿孔板，中间连通。屋顶的几何构造主要与单向弧形的有限空间形成一个平面，可简单通过边缘的折回以保持水平形状得以实现。因为面板重量轻，可通过驳船运输并由船式起重机直接吊装，所以面板尺寸可大。出境大厅的玻璃外立面需性能高、低辐射，以尽量隔热并且允许光线穿透。

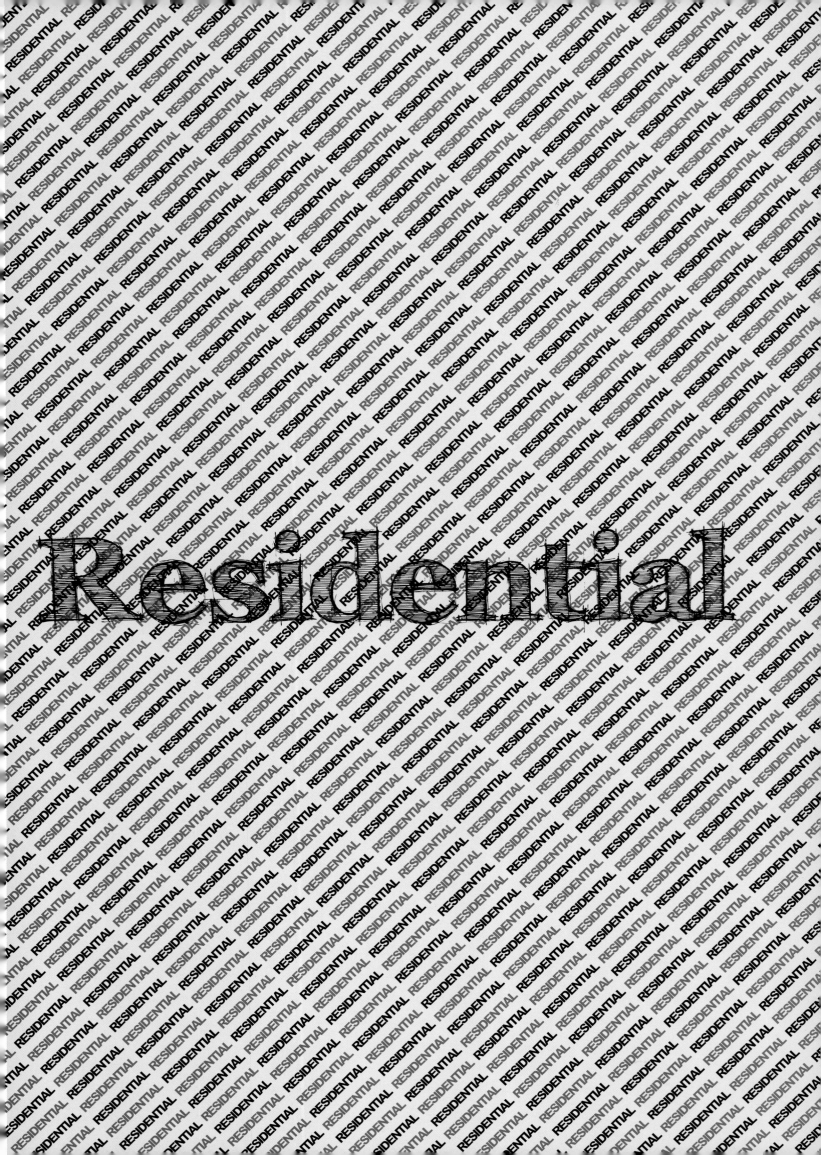
Residential

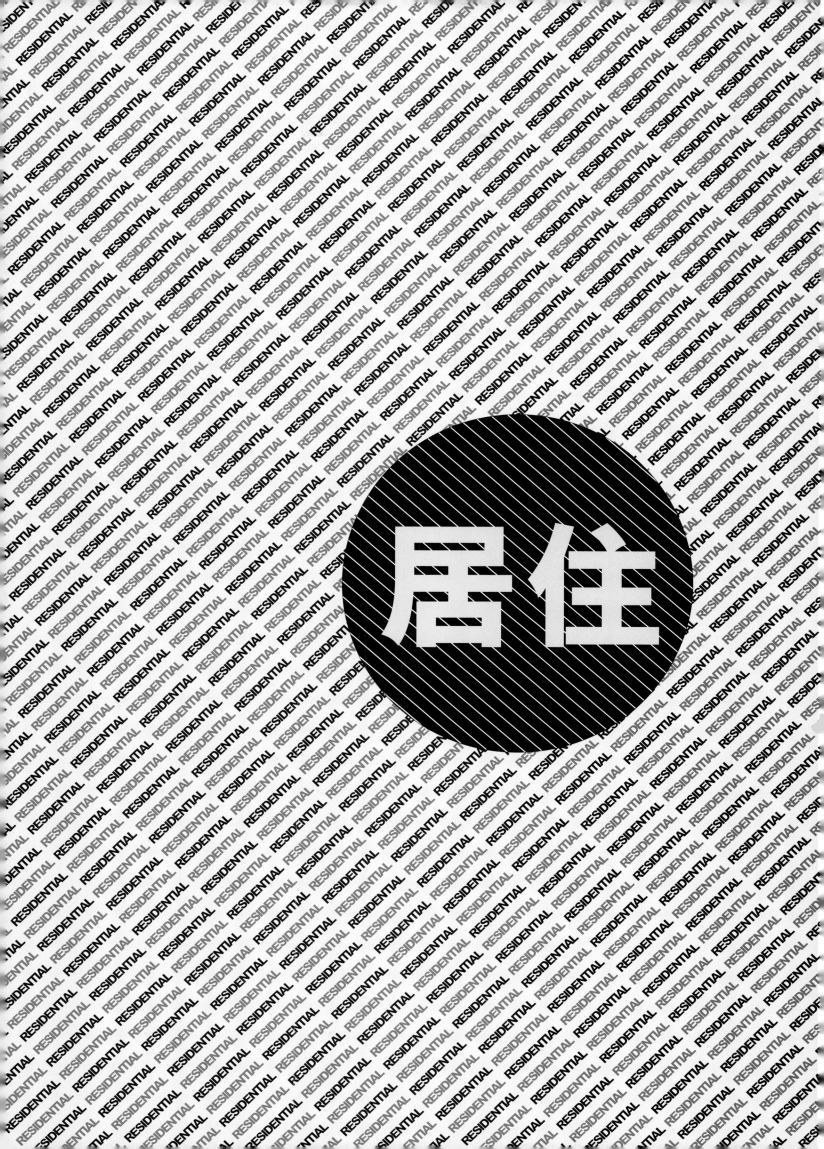

PROBLEM / 问题

1. **LOCAL ENVIRONMENT**
2. **BUILDING STRUCTURE**
3. **FLOW ARCHITECTURE CONCEPTION**
4. **MULTI-FUNCTION**

1. 考虑当地环境

2. 建筑结构

3. 流动性建筑理念

4. 多重功能

1000 Museum
1000 博物馆

Zaha Hadid Architects
works

Zaha Hadid Architects
作品

Architect: Zaha Hadid Architects
Local Architect: O'Donnell Dannwolf Partners
Project Team, Alessio Constantino, Martin Pfleger, Oliver Bray, Theodor Wender, Irena Predalic, Celina Auterio, Carlota Boyer
Client: 1000 Biscayne Tower, LLC
Location: Miami Beach, USA
Gross Area: 84,637 m²

设计公司：扎哈·哈迪德建筑设计公司
本地设计公司：O'Donnell Dannwolf Partners
项目团队成员：Alessio Constantino，Martin Pfleger，Oliver Bray，Theodor Wender，Irena Predalic，Celina Auterio 以及 Carlota Boyer
项目客户：1000 Biscayne Tower, LLC
项目地点：美国迈阿密海滩
总建筑面积：84,637 平方米

DESIGN REQUIREMENTS
设 计 要 求

The project is no less Zaha at this unprecedented scale. A concrete exoskeleton structures the perimeter of the tower in a web of flowing lines that integrates lateral bracing within the lines of structural support. Columnar lines near the base splay out to meet at the corners, forming a rigid tube highly resistant to Miami's demanding wind loads.

项目位于美国迈阿密比斯坎湾大道，正对面是博物馆公园，是欣赏比斯坎湾以及迈阿密海滩风景的绝佳位置。项目的塔楼部分为一个62层，包含83个超级豪华公寓单元的住宅。

THE SOLUTION / 解决方式

① 地基附近的圆柱状线条缓缓展开，最终与建筑大楼的各个角落区域汇合，形成一根根坚固的管状物，抵御来自海洋的强风对迈阿密海滩的侵袭。

② 整个建筑外形像一个连续的流动框架，以避免迈阿密建筑常用的"通用现代主义类型学"。

③ 巨大的塔楼将拥有商业空间，几个用作停车场的楼层以及天空酒廊和水上中心，四个层次的游泳池，日光浴，健身场所和私人停机坪。

④ 建筑师设计出来的曲线线条可以起到对角托架的支撑作用以抵御飓风的侵袭。

Elevation

Details

Details

Project Overview / 项目概况

Design strategies
With structure at the perimeter, the interior floor plates are almost column free, allowing maximum variation in floor plans. The bottom two-thirds of the tower has two units per floor, while the upper third boasts units that occupy the entire floor. The moving, curving lines of the exoskeleton mean that each succeeding floor plan is slightly different from the last. On the lower floors, terraces occupy the corners; on the upper floors, the terraces are tucked in from the edges.

The circular lines slowly extend around the land base, and finally join each area of the building, to form solid cylinders for resisting against the strong wind from the sea towards Miami beach. ❶

The whole building looks like a continuously flowing frame to avoid general modern typology in Miami architecture. ❷

This huge tower will include commercial space, several floors used as parking lots, hanging bar, water centre, four-layer swimming pool, sunlight bath, health club and private airport. ❸

The curves designed may work as a diagonal support to fight against typhoon. ❹

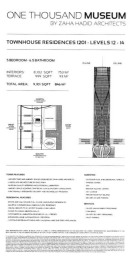
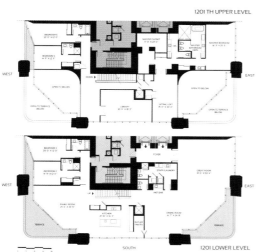
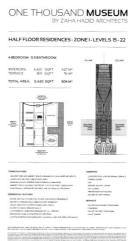
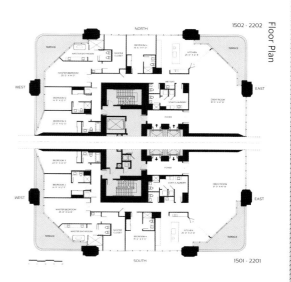

设计策略

虽然对扎哈而言,该项目建筑规模是第一次接触,而相关的建筑设计却完全不失她的一贯个人风格。该摩天大楼外观呈一个混凝土外骨骼结构,采用了网状的流动性线条设计;而结构支撑形成的多条交错线条同时也将整个建筑的横向支撑融合统一成一个整体。地基附近的圆柱状线条缓缓展开,并最终与建筑大楼的各个角落区域汇合,形成一根根坚固的管状物,可抵御来自海洋的强风对迈阿密海滩的侵袭。

Zaha Hadid Architects project director Chris Lepine says that the structure — which appears as if it were eroded from a solid — reads from top to bottom as one continuous liquid frame. The tower represents a line of research in high-rise construction that explores a fluid architectural expression consistent with engineering for the entire height of the structure. The emphasis is on expressing the dynamism of the structure in an integrated whole that avoids the frequent typology of a tower resting on a base.

Lepine points out that while the architects have worked to express the structure and its beauty within a tall, slender tower, the structure itself is "purposeful" in that it is rigid, stiff, and hurricane resistant. It's not a diagrid structure, but its curving lines allow a diagonal bracketing action. "We had this idea of a fluidity that is both structural and architectural," Lepine says.

来自扎哈·哈迪德建筑设计公司的项目总监 Chris Lepine 认为该摩天大楼的结构——整体外形仿佛是一块被侵蚀的磐石,从上往下看时,它又像一个连续的流动框架。该摩天大楼项目代表着建筑师们对高层建筑所做一系列研究的最终结晶,在探索流动性建筑理念的同时,又必须将整个大楼结构的整体高度的工程学原理考虑在内。该建筑设计的重点在于强调整个建筑统一结构的动态性,并同时避免迈阿密建筑常用的"通用现代主义类型学"。
Lepine 指出,虽然建筑师们已经努力在一幢外形纤细的摩天大楼内部构建出优美的建筑结构,但是结构本身还是带有其"目的性"的,因为它必须是坚固的、稳定的并且能抵御飓风侵袭的。它并非是一种斜肋构架结构——但是设计出来的曲线线条可以起到对角托架的支撑作用。Lepine 说:"我们关于流动性建筑设计的想法是基于对结构性和建筑特点的考量而得出的。"

Instead of simply cladding a steel frame, the architects are designing expressive form-work, which can be reused as construction progresses up the tower. The concrete will be painted so that its finished surface is also the architectural finish. "A lot of innovation comes in how we build the form work. We're looking at several solutions," continues Lepine. Behind the exoskeleton, the architects have created a folded, faceted, crystal-like facade to contrast with the solidity of the exoskeleton. The dependable Miami sun will create plays of light on the glass within the structural frame.

"What you see is literally structure getting thicker and thinner, as needed," Lepine says. "There's a continuity between the disciplines, between the architecture and engineering, to create that impression."

建筑师们没有采用不锈钢框架覆层的设计方法，而将设计出一系列形象的建筑用模子材料。在向上建设该栋摩天大楼的较高楼层时，仍然可以重复使用这些材料。该建筑大楼的混凝土表面会涂上一层有色油漆，这一层表面装饰也将自然成为整栋建筑大楼的建筑表面装饰。Lepine 继续说："在建造建筑模子方面，我们提出了很多比较有新意的想法。我们现在正着手处理着几大解决方案。"在外骨骼结构的后面，建筑师们同时也建造了一个类似褶皱结晶体的立面，恰好与外骨骼结构的坚固性形成鲜明的对比。在这一结构框架内，迈阿密充足的阳光总是会如期照在玻璃面上，营造出独特的光影效果。

Lepine 说："从字面意义来看，你所理解的结构要么是根据建筑需要变更厚，要么是变薄。为了营造出那种效果，实际上不同学科之间，以及建筑与工程之间都具有一定的连贯性。"

PROBLEM / 问题

1. **MULTI-FUNCTION**
2. **HIGH DENSITY**
3. **URBAN FABRIC**
4. **STRUCTURE AND BUDGET**

1. 多重混合功能
2. 高密度
3. 城市肌理
4. 结构与预算

Peruri 88
Peruri 88 大厦

MVRDV
works

MVRDV
作品

Architec: MVRDV
Client: Wijiya Karya – Benhill Property, Jakarta, Indonesia
Location: Jakarta, Indonesia
Gross Floor Area: 360,000 m²

设计公司：MVRDV
客户：Wijiya Karya - Benhill 资产公司（印尼雅加达）
地点：印度尼西亚雅加达
总建筑面积：360 000 平方米

DESIGN REQUIREMENTS
设计要求

Former coin factory, the premise is Peruri's property. It's located in Jl. Palatehan 4 district, Jakarta, next to future metro station. Therefore, the project needs to satisfy Jakarta's demands for green space and highly intense space. Besides, it should also respect the existing urban fabric.

场地由 Peruri 所有，之前是铸币厂，坐落在雅加达 Jl. Palatehan 4 地区，现今紧邻城市未来地铁站，项目需满足雅加达对绿色空间的需求与高密度空间的需求，同时尊重现有的城市肌理。

① 建筑包含了建造在商业底座上方的四栋高耸的塔楼,分布其中的包括一系列公寓、酒店、办公场所、店铺、影院、清真寺以及由户外电梯接入的超高露天剧场。

② 整座建筑化整为零,将不同功能体通过叠加,旋转,相互围绕,预留出多处户外空间,供人们纳凉。

③ 项目运用了新鲜、高密度、社会化的绿色迷你城市理念,体现了从自身城市肌理生长出的现代标志。

④ Arup 将会继续开发及合理化建筑的结构,以满足建筑规范和预算。

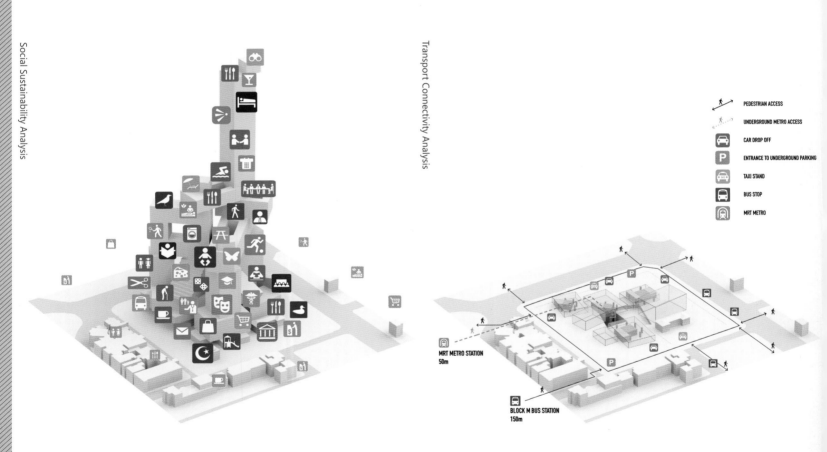

Social Sustainability Analysis

Transport Connectivity Analysis

Project Overview / 项目概况

Location analysis
Peruri 88 combines Jakarta's need for green space with Jakarta's need for higher densities whilst respecting the typologies of the current urban fabric. The site, which is owned by Peruri, is located at Jl. Palatehan 4 Jakarta, a block formerly used as Mint which sits right next to a future metro station.

区位分析
Peruri 88 坐落在雅加达 Jl. Palatehan 4 区,这片土地为 Peruri 所有,原只是个铸币厂。现今它紧邻城市未来的地铁站,不仅将都市绿色空间及高密度需求紧密结合,同时也充分尊重城市现有肌理。

The building comprises four slender towers built on the commercial base, housing apartments, hotel, offices, shops, cinema, mosque and super-high open theatre connected to the outdoor lift. ❶

The whole building is to leave many outdoor spaces by stacking, rotating and revolving different functional areas, for residence to enjoy the cool. ❷

The project adopts novel, high-density and socialised green mini-urban conception, presenting the modernity inspired from the existing urban fabric. ❸

Arup will continue to develop and rationalise the building structure to meet building standards and budget requirements. ❹

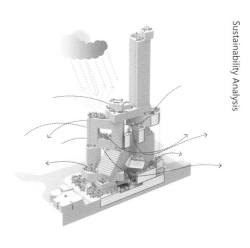

Sustainability Analysis

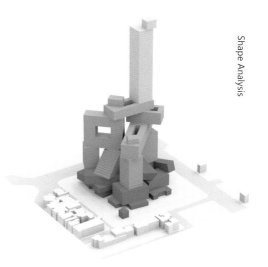

Shape Analysis

The building's structure has five principal cores and is less complex than visually apparent. Four traditional constructed tall towers rise up between which bridging floors will be constructed. Arup will continue to develop and rationalise the structure to satisfy regulations and the budget.

项目概况

建筑结构由五个主要部分组成，其实视觉上看起来并没有那么复杂。四座传统结构的高楼由连接层嫁接，并将其围绕，相互交错。奥雅纳也会继续对建筑结构进行开发，并将其合理化以满足建筑规范及预算要求。

Design strategies

The mixed-use project offers a great variety of office and housing typologies, from large office surfaces to living/working units, from lofts to townhouses, from terraced houses to patio living. Each of these stacked urban blocks comes with a semi-public roof park, an abundance of gardens, playgrounds, spas, gyms outdoor restaurants and swimming pools available to the inhabitants and office employees. The tall trees on these decks will provide extra shade whilst the height of the parks allows for a cooling breeze. The high rise, a luxury hotel from the 44th floor to the 86th floor, rises from a platform with park, swimming pool and the marriage house. On top of the hotel a panoramic restaurant and viewing platform complete the structure at the 88th floor.

设计策略

该综合型项目提供各种各样的办公空间及住宅户型，大型办公区或住宅或工作单位，阁楼或排屋，连栋房屋或庭院。建筑体堆叠如山，每个建筑体设有一处半开放的屋顶公园、大面积的花园、游乐场、水疗中心、健身、室外餐厅及游泳池，居民和员工可以在这里放松自我，舒筋散骨。观景台高树林立，为人们多带来一片纳凉之地。高处的公园可谓凉风习习，让人流连忘返。大厦从44到86层皆属于豪华酒店，仿佛从某一平台拔地而起。平台上花园、游泳池、婚礼礼堂样样不缺。酒店顶层的全景观餐厅和观景台给88层楼的布局划上了完美的句号。

The commercial podium which is located from levels B2 to the 7th floor is designed by Jerde Partnership with MVRDV. Its most characteristic feature is the central plaza, sheltered by the stacked volumes of the mid-rise, offering multiple outdoor layers of restaurants and shadow and natural ventilation. A series of escalators connects the shopping and retail centre to the parks of the mid-rise.

The Peruri 88 commercial podium reflects the city's historic islands with reflective bodies of water and landscape traversing the public street levels, while integrating a sunken garden plaza.

商业裙楼始于地下 2 层，终于地上 7 层。商业裙楼由捷得建筑合伙事务所和 MVRDV 携手设计而成。其中最能表现出该建筑特色的是中央广场。中层建筑如群山四面环绕，中央广场隐秘其中。商业裙楼提供了多层的户外餐厅，户外纳凉地，让人们感受自然风的凉爽。一系列扶梯串联起商场、购物中心及中层花园，交通极为便捷。
Peruri 88 商业裙楼通过水反射光线的物理特性以及贯穿各层公共街道的景观，再结合下沉花园广场，向世人表达了岛国城市久远的历史。

PROBLEM / 问题

① **LANDMARK**
② **MULTI-FUNCTION**
③ **INTEGRATION WITH SURROUNDING ENVIRONMENT**

1. 地标性
2. 多功能混合
3. 融入周围环境

Dominique Perrault
the Blade

Dominique perrault
the blade

Dominique Perrault
works
Dominique perrault
作品

Architect: Dominique perrault
Client: Dreamhub-Yongsan Developmet CO, Ltd
Location: seoul, korea
Surface Area: 131,700 m²
Function: Business forum, wellness lobby, offices and panorama lobby

设计公司：Dominique perrault
客户：dreamhub - yongsan development co., ltd.
地点：韩国首尔
地上面积：131 700 平方米
功能：商业论坛、健身中心、大堂、办公

DESIGN REQUIREMENTS
设 计 要 求

The project is located in Yongsan International Business District, Seoul designed by Daniel Libeskind.

项目为位于由丹尼尔·里伯斯金汉城总体规划设计的首尔龙山国际商务区

THE SOLUTION 解决方式

① 1. 项目大厦通过独特的设计手法，整体上仿佛就是一面菱形棱镜，呈现出来的动态轻盈美将使其成为该地区的地标性建筑。

② 项目除了设置了相应的办公区之外，该项目还集住宅、商铺以及众多政府设施（文化设施、教育以及交通基础设施）于一体。

③ 整栋大厦仿佛是一台光学仪器，各个立面的格局被打碎，然后又与周边景观重新组合在一起，形成了一道全新的风景。

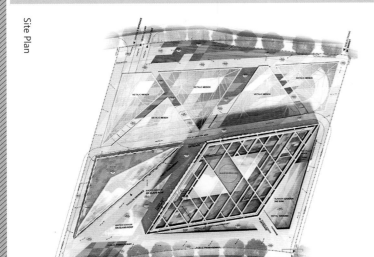

Site Plan

Location Plan

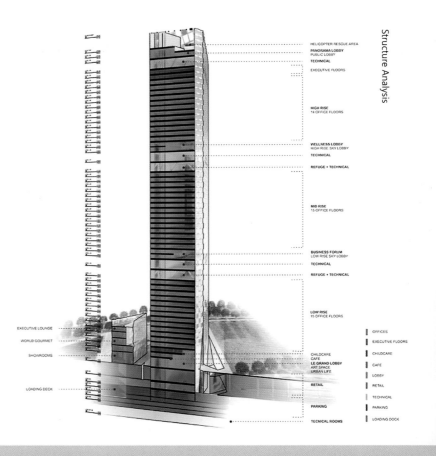

Structure Analysis

Project Overview / 项目概况

Project overview
The Yongsan International Business centre, ambitious programme of nearly 3 millions square metres, is organised as an archipelago of vertical buildings inter-connected a by large park.

项目概况
龙山国际商业区项目规模宏大，占地面积近三百万平方米，一个大型公园将一座座笔直的大楼相互连接于一体。

The building after unique design looks like a piece of rhombus mirror on the whole, presenting a dynamic beauty that earns it the title of landmark in that area. ①

Except offices, the building also houses residences, shops and several governmental facilities (cultural facility, education and transportation facility as well). ②

The whole building looks like an optical instrument, with all elevations broken and combined with the surroundings again to create a totally new landscape. ③

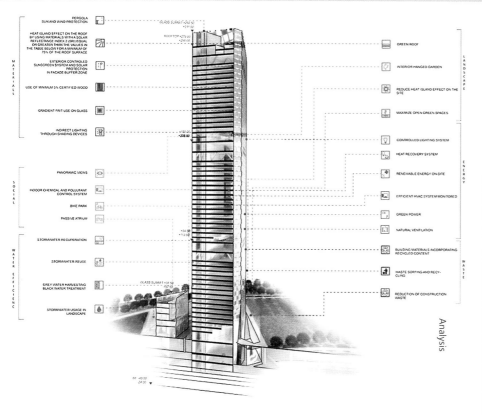

Analysis

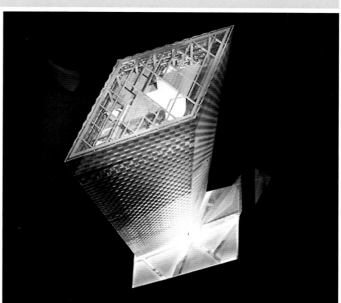

Connected to three other major business centres of the city, the future Business centre is developed away from the large monofunctional complexes, offering beyond the offices areas, housing, shops and many government facilities (cultural facilities, education and transport infrastructure).

这一未来的商业中心将与其所在城市的其他三个商业中心连接在一起。但是该项目却未沿用单一功能大规模城市综合体的传统设计方法。除了设置了相应的办公区之外，该项目还集住宅、商铺以及众多政府设施（文化设施、教育以及交通基础设施）于一体。

Design concept

By its silhouette and its dynamic allure, the tower establishes itself in the area as a geographical landmark. Its mysterious shape appears like a totem, an iconic figure. It is not a square or a round building, but a rhomboid prism, arranged in a way that makes it look different depending on the angle of approach. Inspired by its slender shape and sharp edges, the tower has been named The Blade.

设计理念

Dominique Perrault 是唯一受邀的法国建筑师,再次参与了首尔的变革。在梨花女子大学完成学业之后,这位建筑师通过独特的建筑风格,参与到了对未来商务区的定义。

该项目大厦轮廓分明,呈现出来的动态轻盈美将使其成为该地区的地标性建筑。该项目大厦神秘的外形轮廓似某个部落的图腾,又似某种拟人化的图形。

Design strategies

In the way of a sheath, the skin of the tower is clad with glass, reflecting light and its environment, thus releasing a luminous halo which envelopes the silhouette of the tower. This vibration of the building's skin appears and disappears according to the viewing angle, creating a living architecture, transforming itself with the movements of the sun and the changes of light.

The project sculptures the void like a luxurious material, offering space, light and views of the grand Seoul landscape. The Grand Lobby, the Business Forum, the Wellness or the Panorama Lobby constitute as many cut-outs in the tower volume, dedicated to promenades and relaxation. This superposition of voids contrasts with the constructed volume of adjacent towers and accentuates the lightness of the tower prism.

The voids offer respirations and accommodate collective spaces open to the landscape. At night, they dematerialise the silhouette of the tower, which appears then like a precious carved stone.

设计策略

正如刀锋需要刀鞘的保护一样，该项目大厦的表皮覆盖有一层可反射光线和周边环境的玻璃材料。在阳光的照射下，经玻璃的反射作用，整栋大厦的轮廓皆笼罩在其形成的光环之中。根据观察角度的不同，该建筑表皮时而呈现出动态之美，时而又呈现出静态之优雅。整栋大厦仿佛"活"了过来，会随着阳光的移动和光影的瞬时变幻而不断改变其本身的外在造型。
该项目大厦上设计的孔洞是采用雕刻的方式，像对待一件豪华材料一样进行装饰的，不仅能营造出足够的空间和充足的光线条件，还能为欣赏宏伟的首尔城市风景提供绝佳的位置地点。大堂、商业论坛区、健身中心或者全景大厅就像是一幅幅镶嵌在该项目大厦体量上的剪影画，专供人们漫步和休憩使用。错位排列的孔洞与邻近各大高楼组合构造而成的体量形成了鲜明的对比，并突出了该项目大厦如棱镜般的晶莹剔透之美。
该项目大厦的孔洞设计不仅能为建筑通风创造条件，而且还有益于集体空间与周围景观的开放式融合。夜晚时分，这些孔洞设计将搅乱整栋大楼的外形轮廓，因为夜晚时的整栋大楼就像是一块经精心雕刻的巨石。

Within the effervescence of the emerging architectural styles, The Blade contrasts by being rooted in the urban reality, in a dialogue of light and reflections with the neighbouring towers. Like an optical instrument, its facade fragments and then reconstructs the neighbouring landscape to create a new one.

该项目大厦结构既不成正方形，亦不成圆形，大楼整体上仿佛就是一面菱形棱镜。通过独特的设计手法，身处于不同的观察角度，大楼所呈现出的外形轮廓也略有不同。受其纤细的外形轮廓和尖锐分明的边缘线条启发，该项目大厦已被命名为"尖锋大厦"。
在各种新兴建筑风格备受推崇之际，尖锋大厦在整个城市背景的映衬下显得尤其突出。在阳光的照射下，该项目大厦与其附近的各大高楼交相辉映。整栋大厦仿佛是一台光学仪器，各个立面的格局被打碎，然后又与周边景观重新组合在一起，形成了一道全新的风景。

PROBLEM / 问题

1) SHARE AND COMMUNICATION SPACE
2) USER DEMANDS
3) SOCIALITY
4) ECONOMY

1. 共享与交流空间
2. 用户需求
3. 社会性
4. 经济性

Labitzke-Mitte
LABITZKE -MITTE

gus wustemann architect
works
gus wustemann architect
作品

Architect: gus wustemann architect
Location: Zurich, Switzerland

设计公司：gus wustemann architect
地点：瑞士苏黎世

DESIGN REQUIREMENTS
设 计 要 求

The goal is to create a centre in periphery, not only a building, but also a space, an urban cluster, a vertical line connected to the city or a horizontal line to the district. Besides, it also serves as a social centre, to satisfy the general demands from an urban community.

我们的目标是在外围创造一个中心——不仅是作为一项建筑特色，并同时作为一个空间，作为城市综合群组、作为与城市联系一体的垂直直线或者与地区联系一体的水平直线。它同时也是一个社会中心，能够考虑一个城市社区的普遍需求。

THE SOLUTION 解决方式

① 建筑大楼的第一层将用作多变、透明、风景优美以及营造氛围的总空间，而且相互联系的程度也非常高。

② 设有共享烧烤区的半开放式屋顶花园以及特别配备有屋顶露台的公共空间将成为建筑使用者们聚会的场所。

③ 整片区域通过绿色景观，水景和社区广场为城市居民提供舒适的居住环境。

④ 实用主义的原则，结构和材料的使用以实用性为原则，在具备技术性美感的同时营造出独特的氛围。

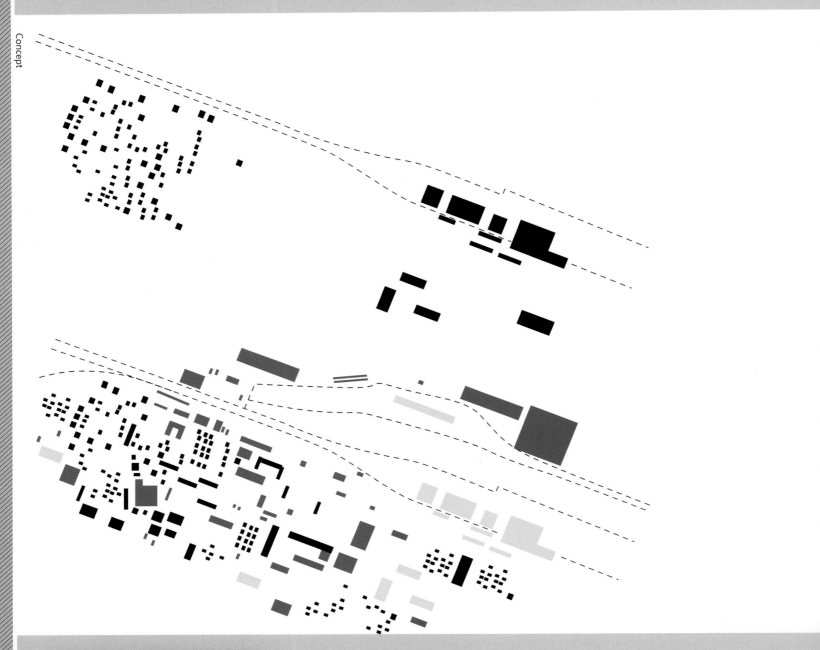

Project Overview / 项目概况

Location analysis
Labitzke-Mitte, a district with identity: new urban structure which differs geographically, topographically, historically, socially or mentally von anderen. In urban transition agglomeration

区位分析
Labitzke-Mitte 是一片特点较为突出的区域：新的城市结构无论从地理学、地形学、历史学、社会学甚至精神学的角度而言都与众不同。在城市转型聚集（聚集：疏松岩石碎片的累积）的过程中，某个"地区"会为自己定位，并在情感

The first floor of the building will be designed to be a changing, transparent and picturesque space that creates pleasant atmosphere and shares close connection between each different part.

The public space like semi-open roofed garden with BBQ function and roofed balcony specially designed will turn out a best choice for users' party.

The entire area, through green landscape, water views and community square, provides a comfortable residence for the citizens.

Under the principle of practicality, the structure and materials are used to create unique atmosphere with technically beauty.

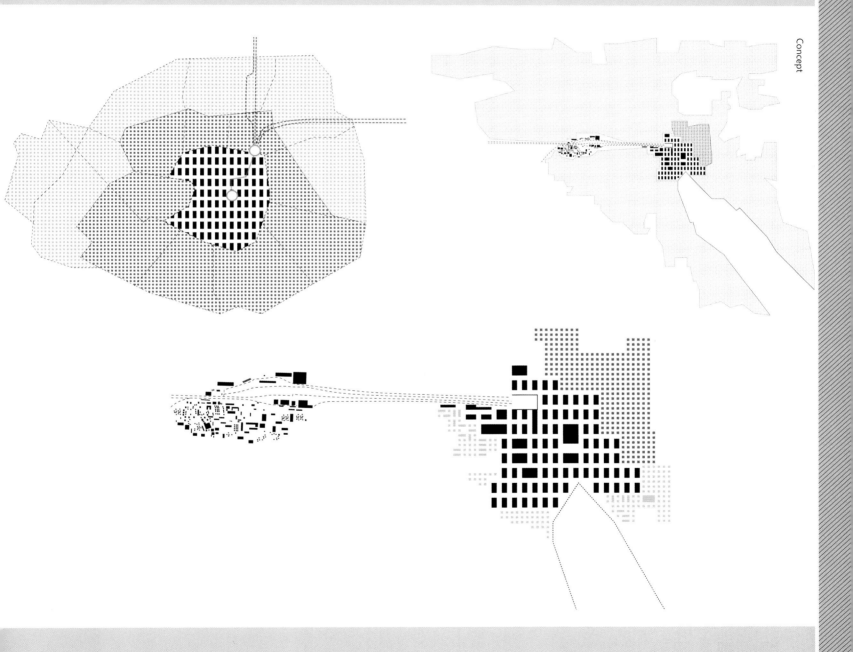

Concept

(agglomerate: accumulation of loose rock debris)means a "quarter" to constitute itself as orientation with emotionally comprehensible size (identity), With the unaligned, indefinite, accidental openness and transparency, one direction is given, without, however, sacrificing optionality.

上演变成可让人理解的规模（特点）。然而，当给出一个随机、不确定甚至是偶然的方向时，它又会在不牺牲可选择性的情况下，兼具开放和透明的特点。

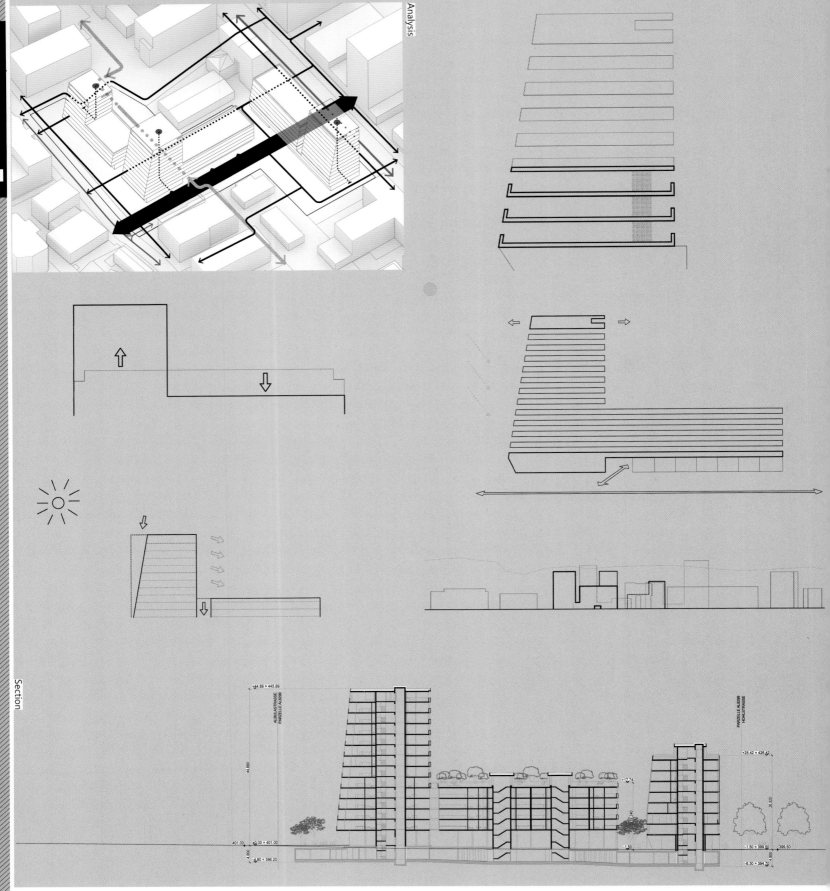

Design concept
The goal is to create a centre in the peripheral - not only as an architectural character, but as a space, as urban-integrated group, as city-related vertical lines, horizontal district-related.

设计理念
我们的目标是在外围创造一个中心——不仅是作为一项建筑特色,并同时作为一个空间,作为城市综合群组、作为与城市联系一体的垂直直线或者与地区联系一体的水平直线。它同时也是一个社会中心,能够考虑一个城市社区的普遍需求。

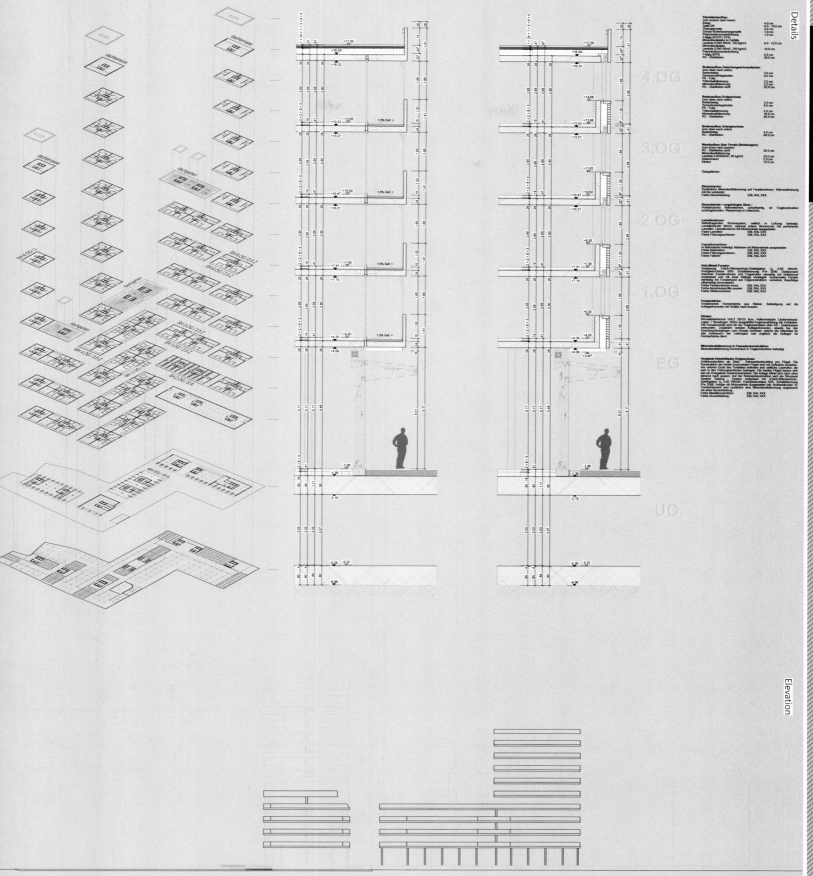

Also as a social centre, consider common needs for an urban community are considered. The range and networking provides an urban centre, almost from the periphery to the centre. The daycare is the living heart of the centre, where elderly residents will be attracted by the children. For this purpose, a small town grocery store hybrid would be ideal.

几乎从外围到中心区域,社区配备的设施范围和体系都符合城市中心的生活要求。日托所是该社区中心非常关键的一部分,因为孩子们生活在这里,老年人的生活也会因孩子们嬉闹的场景而变得丰富许多。基于此目的,再另外设置一家小型城镇杂货铺将会是个不错的选择。

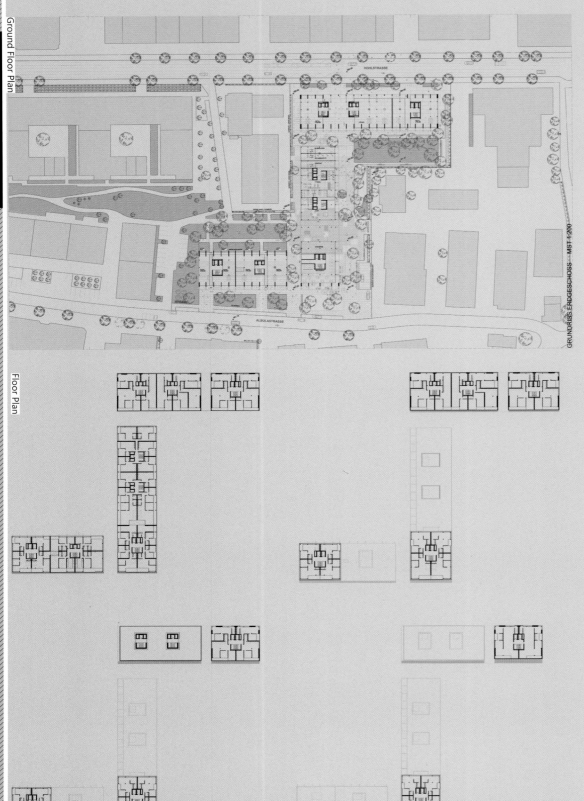

Design strategies

Interior / exterior space, free space and surroundings Dissolve the dichotomy of inside / outside as an either / or in transition: The ground floor is used as a variable transparent, scenic, atmospheric total space understood with a high degree of crosslinking. The ground floor will also be children's play areas and niches for the level of encounter in the neighbourhood. In the complex itself the semi-public roof gardens with shared barbecue areas and especially the common room with a roof terrace with views of the neighbourhood hybrid are intended as a meeting place. Space: Big respect to the centre, several playgrounds and rock climbing as a meeting room; children are the reference point.

Meadows, trees and Albula Place: The existing landscaped public courtyard is continued up to and through the middle. The immediately adjoining the centre outdoor spaces are always in-touch with the centre. The ground floor is to the covered in mic summer as outdoor space. Green walls close off this flowing, networked outer space. Trees over the whole area distributed shade and are important reference points for the residents as nature in the city. The drink fountain is an important urban element, where children play in the summer. Two large green lawns are soft places to play and relax. At the Albulastrasse a neighbourhood square is created along with a front garden and a café bar with evening sun terrace.

设计策略

内／外部空间、自由空间以及周边环境：在转型过渡的过程中，消除内部与外部之间的分界线。建筑大楼的第一层将用作多变、透明、风景优美以及营造氛围的总空间，而且相互联系的程度也非常高。建筑大楼第一层不仅将成为孩子们的游乐场所，而且也将会是住户与邻居们偶然相遇的空间。就这栋建筑综合体本身而言，设有共享烧烤区的半开放式屋顶花园以及特别配备有屋顶露台（在这里可以欣赏周边其他建筑综合体的风景）的公共空间将成为建筑使用者们聚会的场所。空间：需与建筑中心、几大游乐场以及攀岩运动场所形成相呼应的关系——孩子们是主要的参考点。

草地，树木与阿尔布拉广场。现有的景观公共庭院一直向上延伸并贯穿了建筑物的中间部分。直接与该中心毗连的户外空间总是与中心联系紧密。盛夏时分，建筑大楼的第一层将转变为有屋顶覆盖的户外空间。绿墙隔绝了这一流动的、呈网络状分布的外部空间。整片区域内都种满了树木，随时可提供遮阳和纳凉场所，这也是城市居民选择在这里居住的重要参考点。人造喷泉是一个重要的城市生活元素。zürcher 水景区可供孩子们可在夏天放学的傍晚尽情嬉戏。两片大型的绿色草坪区温馨而柔软，是人们娱乐和放松的好去处。在 Albulastrasse，设有一个社区广场，而在该广场的前方是一个花园和一家咖啡酒吧，站在这里的露台上，可尽情欣赏落日的余晖。

Address Education hollow road : along the hollow road is the town of hybrid opposite in dialogue with the high points and marks a clear address; a highly structured urban facade that opens further on the upper floors. On the ground floor commercial programs or services are provided, a generous passage is the entrance to Labitzke CENTRE.
Construction, materials and aesthetics of the Functional Living, Pragmatic: clear practical structure, honest Konstruktions¬und Cover material: speaking materials, concise quality, technical beauty, sensuality. Economy: reduction to the essentials. Not simulate the industrialists, but take up as identity, get, make them visible. A support structure with inner, load-bearing concrete skeleton allows a flexible, economically sensible dimensioning of the supporting elements. The outer wall is a bivalve exposed brick construction, the interior walls are plastered walls with gypsum plaster caster. Following the serious industrial architecture of the area, the Swiss tradition of craftsmanship and architecture, we use only raw materials, exposed concrete ceilings and bright exposed brickwork give structure and atmosphere. Raw surfaces lend authenticity.

Address Education 凹路：沿着这条凹路一直走下去，便可以到达该建筑综合体所在的城镇（对面是与之相呼应的较高地点，布局清晰明确）；高度结构化的城市立面进一步向建筑物的上部楼层延伸。商业功能或者服务功能皆设置在第一层；经由一条开阔的走道可直接到达 Labitzke 中心。
是否回归真实和本源是鉴别无所不有以及无所不在的新一代人的关键点。从何物存在，到存在何方，再到如何存在？原材料是无法进行
建造物、材料以及功能性和实用主义生活的美感：清晰的实用性结构，可靠的 Konstruktions-und。覆盖材料：就材料而言，它必须具有简约的特点，具备技术性美感同时还能带来感官享受。经济性：仅留下最精华的部分，其余缩减。
不冒充工业主义者，但是它必须具有自己的特点，而且这些特点必须显而易见。支撑结构的内部采用的是钢筋混凝土骨架，以便对各大需支撑元素进行灵活而又在经济上比较合理的尺寸计算。外墙采用的是露天的两瓣式砖石建筑结构，而内墙则涂上了一层厚厚的石膏灰泥。为了延续该地区工业建筑的严肃风格以及继承瑞士的传统技艺和建筑手法，我们只使用原材料。天花板上裸露的混凝土以及直接暴露在外鲜亮的砖砌造型更能让人体会到建筑的结构感和营造出的独特氛围。原始表面更具真实性特点。

The return to authenticity and origin are important points of identification of a new generation who have everything and were everywhere. From what there is, where does, how is it done? Raw material is non-hierarchical and non-social contaminated, it leaves scope for personal development and interpretation.

The Living with south facing terraced and high points is especially designed for the exposure and the relation to outer space. Various floor plans reflect a modern conception of urban living, no hierarchy, raw materials and program clears convey sensuality and security, at the same time with great views of the city, and track the Üetliberg. The south facades are glass facades with generous private outdoor spaces, and CENTRE neighborhood hybrid are designed with 50% by weight and 50% glass so that a maximum views and exposure is achieved with minimal insight in the urban context. The city has an urban hybrid, massive facade of the hollow road and a softer more open facade to the CENTRE in the south.

Possible stages of construction, we provide the first city to develop hybrid and the hollow road, and then also gradually mid - to provide accommodation and hybrid ready. As architects, we see a possible division in the city CENTRE -and hybrid and hybrid quarters.

等级划分的，同时也是未受社会污染的，它会留有个人发展以及理解想象的空间。
我们还特别设计了带阳台的面南生活空间和一些较高点，从而更好地与户外空间产生联系。各种楼层布局设计都反映了现代都市生活的理念：无等级划分、使用原材料、建筑物的形式表达必须能给人带来安全感和感官享受；同时，还必须具备欣赏城市美丽风景以及特利山景的条件。建筑的南立面使用的是玻璃材料，同时还营造出开阔的私人户外空间。该中心社区建筑综合体 50% 的重量来源于玻璃结构。如此一来，可最大化视野范围和暴光程度，但是对城市肌理的理解程度却最低。这座城市不但有城市建筑综合体、凹路的大规模立面，而且在南边还有朝向中心位置的更加开放、柔和的外立面。
可能的建设阶段，我们准备在第一个城市开发建筑综合体并进行凹路设计；然后逐渐地向建筑中期阶段发展——提供住宅以及成熟的建筑综合体。作为建筑师，我们看到划分城市中心的可能性——即混合性建筑综合体。

PROBLEM / 问题

① SUSTAINABILITY
② DENSE AND COMFORTABLE RESIDENTIAL SPACE
③ LANDSCAPE VIEWS

1. 可持续性
2. 高密度和舒适的住宅空间
3. 景观视野

Pentomonium Towers
Pentomonium 双子塔

Murphy/Jahn
works
Murphy/Jahn 事务所
作品

Architect: Murphy/Jahn
Client: Dreamhub
Location: seoul, korea
Function: Residence

设计公司：墨菲/扬事务所
客户：DreamHub
地点：韩国首尔
功能：住宅

DESIGN REQUIREMENTS
设 计 要 求

This is a high-end residential building developed on the west of Yongsan International Business District, Seoul, Korean.

这是为韩国首尔龙山国际商务区西侧而建造的高端住宅大厦。

① 可打开的玻璃窗扇允许手动打开，以调节室内微气候。

② 包含的隐秘空间可以给提供住户只有在单个别墅中才可以享受到的宝贵城市生活体验。

③ 楼梯和电梯核心位于每个塔楼的东北角落，给住户提供了朝着汉江的视野。

Sketch

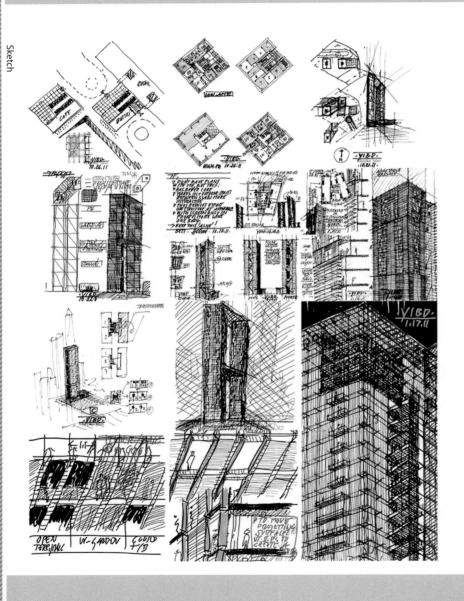

总平面图
Site Plan

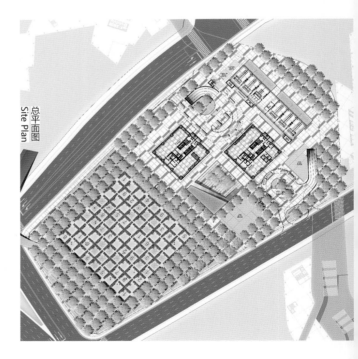

Design strategies

Reaching 320 metres tall, the slender profile emerges from a small square footprint, minimising the number of units on each level while maximising seclusion offering dwellers an urban living experience previously only found with individual homes. The scenic southwestern views are enhanced by internally shifting the vertical circulation cores towards the northeastern elevation. The resulting U-shaped configuration may be subdivided, maintaining their outward perspectives while penthouses use the entire floor.

设计策略

高320米，细长的外形从一个小的正方形基地拔地而起，将每一层的房间最小化，同时，将包含的隐秘空间最大化以提供住户只有在单个别墅中才可以享受到的宝贵城市生活体验。通过把内部的垂直流通核心转移至东北面，从而进一步增强了西南方向的景观。这样，所形成的U形外形可被划分开来，保持向外的视角，同时顶层公寓的住户可以享受整个楼层的面积。

1. The glass window allows manual operation to adjust indoor climate.

2. The private space provides residents with precious urban life experience that can only be enjoyed in villas.

3. The core of stairs and lifts is in the northeast corner of each tower, providing a broad view towards Han River.

The enclosure then shifts between the neighbouring residences, forming bay windows, balconies and winter gardens. At the perimeter, a screen of vertical and horizontal bars within a five-metre-tall by three-meter-wide modular grid aligns the geometries between each of the buildings. The exterior surface serves as a structural framework, providing shade and additional privacy. Four-storey-high parks are carved into the sides at various heights, offering residents recreational areas and lounges. Operable glass panels allow for manual adjustments to the interior climate.

围护结构则在相邻住宅之间转换，行成飘窗、阳台和阳光室。在边缘，一排垂直和水平的杆子在一个5米高3米宽的模数网格中，使每个塔楼的几何图形一致。外立面作为一个结构框架，提供阴凉之处和额外的隐私空间。4层高的公园依靠在不同高度的每一个建筑面上，给住户提供一个娱乐和休闲的处所。可打开的玻璃窗扇允许手动打开，以调节室内微气候。

PROBLEM / 问题

1. **MINIMISING THE ECOLOGICAL FOOTPRINT OF THE BUILDING**
2. **INCREASING GREEN COVER TO MAITAIN SITE BIODIVERSITY**
3. **CAPITALISE ON THE SITE POTENTIAL**

1. 减少建筑物的生态足迹
2. 增加绿地覆盖，以保持现场的生物多样性
3. 利用现场的潜力

Kings House Apartments
国王公寓

The Purple Ink Studio
works
THE PURPLE INK STUDIO
作品

Architect: The Purple Ink Studio
Client: Kini Family, Dubai
Location: Bangalore, India
Area: 743.22 m²
Function: Residential apartments

设计公司：The purple ink studio
客户：　 Kini Family, Dubai
地点：　 印度班加罗尔

DESIGN REQUIREMENTS
设 计 要 求

Occupying a strategic position in the heart of Bangalore, the clients saw a potential for limited "Sky Villas" where architecture would re-interpret new levels of luxury living by capitalising on the potential of the site without compromising on the existing greens.

位于班加罗尔的中心地带,占据战略位置;充分利用场地条件,无需破坏现有绿化,建筑师可以重新阐释豪华住宅的新高度,而客户也发现了有限"天空别墅"的潜力。

① 总规划必须检查并组织用地方案,以解决在该场地环境中的绿色参数下实现超豪华生活的复杂性。

② 每一层楼板将延伸至绿植和阳台中,以给场地环境中创造巨大的多样性。

③ 建筑的功能就像一个选择性的环境过滤器,增强区域环境中的最好因素以满足结构中供暖、制冷和通风的需求。

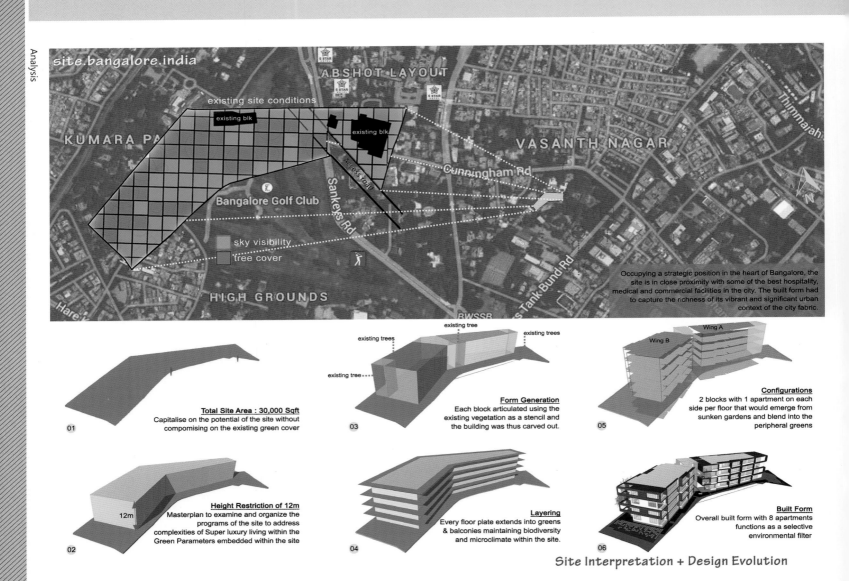

Site Interpretation + Design Evolution

Project Overview / 项目概况

The King's House occupies a strategic location in the heart of Bangalore city, but the site is a quiet hideout from the hustle-bustle of its surroundings. It is in close proximity with some of the best hospitality, medical and commercial facilities in the city. The client's brief saw a potential for limited "Sky Villas" where the architecture would re-interpret new levels of luxury in living.

国王公寓占地位置为班加罗尔市中心,是一个具有战略意义的位置,但又从其周围环境的熙熙攘攘中隐匿起来,十分宁静。项目离城市中最好的服务,医疗和商业设施都很近。客户可以看出一个受限的"天空别墅"的潜力,其建筑风格可重新诠释奢华生活的新层次。

Every floor plate extends into greens and balconies generating great diversity within the site context. ①

The building functions as a selective environmental filter, enhancing the best components of the regional climate to address heating, cooling and ventilation needs of the structure. ②

Daylight analysis

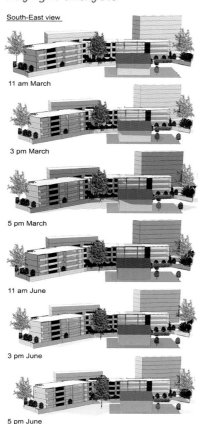

South-East view

11 am March

3 pm March

5 pm March

11 am June

3 pm June

5 pm June

Conclusions from shade analysis : The building Form, the Exterior boxing design and the overall planning were re-generated to suit the design developed from the shade analysis that was carried out. This Reduces the Cooling energy load to the maximum extent.

Green distribution

100 nos. Proposed trees
25 nos. Trees existed before construction

GREEN INSERTIONS

The distribution of greens is a response to the functions in close proximity. The site has tall buffer planting all around with tree heights based on SUN DIRECTIONS and WIND PATTERNS. Areas with private spaces are screened with tall trees and additional buffering with planter beds.

All the planting is planned in THREE LAYERS i.e. trees, shrubs and ground covers resembling a NATURAL HABITAT thus maintaining the BIODIVERSITY of the site and helps in maintaining the site MICROCLIMATE. This vegetation patterns also aids in curbing soil erosion and balancing the soil ecosystem.

TREE - COURT adding private green space at every level

BIOWALL (vertical greens) reduces heat penetration + enhances site biodiversity

PATHWAY with green inserts allows water percolation + ground water recharge

Sustainable responses

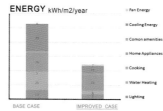

59% of ENERGY is saved by SUSTAINABLE designing i.e Rs.4.0 lakhs / year

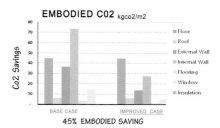

45% EMBODIED SAVING

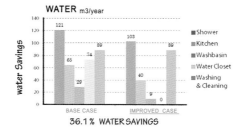

36.1 % WATER SAVINGS

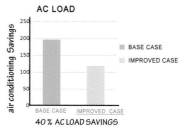

40 % AC LOAD SAVINGS

The design challenge was to capitalise on the potential of the 30,000 sqft site without compomising on the existing green cover and minimising the ecological footprint of the structure. An integrated design approach was followed to evaluate and maximise the energy reductions of the building.

The masterplan had to thus examine and organise the programmes of the site to address the complexities of super luxury living within the Green Parameters embedded within the site context.

这个设计的挑战就是在丝毫不能影响现有绿化的情况下挖掘这三万英尺的场地的潜力，并且将生态足迹减小到最小化。该项目遵循着一个综合的设计方法，以评估建筑节能并将其最大化。
总规划因此必须检查并组织用地方案，以解决在该场地环境中的绿色参数下实现超豪华生活的复杂性。

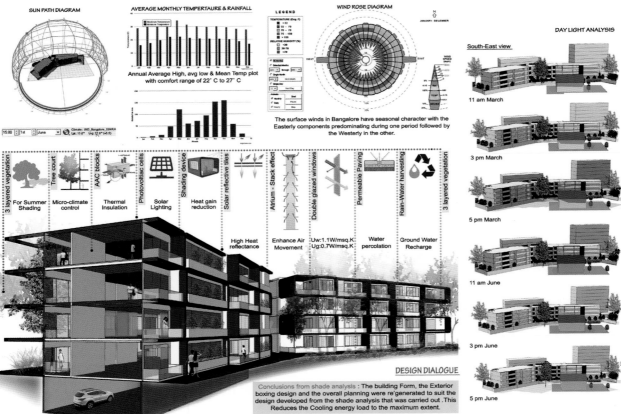

Facade generation - responsive to climatology and function

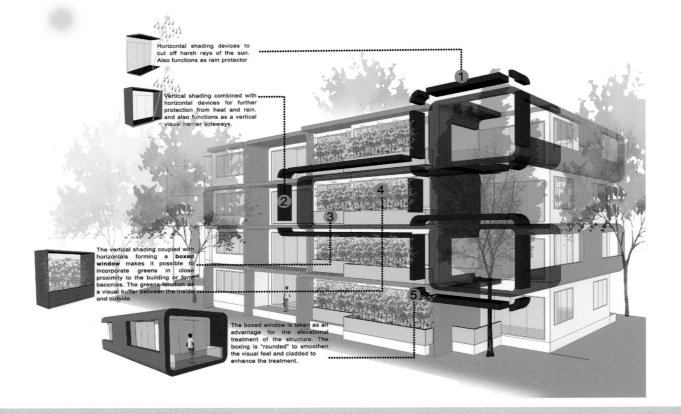

Design narrative

The idea of going higher was restricted by the existing byelaws which did not allow the building to go beyond 12m. The masterplan had to thus examine and organise the programmes of the site to address the complexities of super luxury living within the Green Parameters embedded within the site context. Two blocks were planned to house one apartment on each side per floor that would emerge from sunken gardens and blend into the peripheral greens amidst the site. Each block was articulated using the existing vegetation as a stencil

设计叙述

建筑向高处发展的想法受到了现有规章制度的限制，规定建筑不得超过12米。总规划因此必须去调研并组织用地方案，以解决在该现有基地环境内的绿色参数下实现超豪华生活的复杂性。计划两个裙楼每一层的两侧均有一套公寓，从

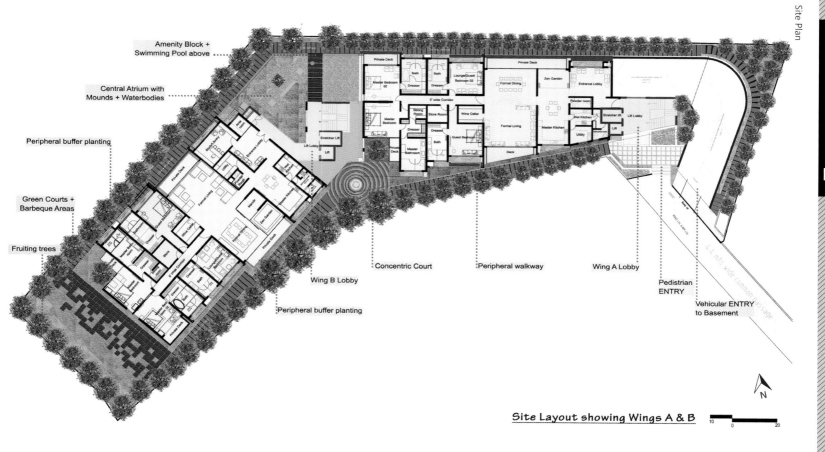

Site Plan

- Amenity Block + Swimming Pool above
- Central Atrium with Mounds + Waterbodies
- Peripheral buffer planting
- Green Courts + Barbeque Areas
- Fruiting trees
- Peripheral buffer planting
- Concentric Court
- Wing B Lobby
- Peripheral walkway
- Wing A Lobby
- Pedestrian ENTRY
- Vehicular ENTRY to Basement

Site Layout showing Wings A & B

and building was thus carved out. To compensate on the loss of lower vegetation from the site during the construction, every floor plate extends into greens and balconies generating great diversity within the site context.

下沉花园中突出，然后融入场地中周边的绿色环境中去。每一个街区都利用现有植被作为漏印板连接起来，如此一来，建筑就凸显出来了。为补偿施工期间场地之下的植被损失，每一层楼板将延伸至绿植和阳台中，以在场地环境中创造巨大的多样性。

Bioclimatic design

An integrated design approach was followed to evaluate and maximise the energy reductions of the building. Solar studies and simulations were used to generate data regarding daylighting, shadow analysis, rainfall pattern and shading systems. These studies, along with lighting analysis, were critical to generate the load calculations and sizing and selection of all the Mechanical Systems. To optimise the cooling effect, the building mass and window openings were shaped and sized to best capture the breezes based on Computer Generated Simulations. The Vertical Shading devices combined with Horizontals cut off harsh rays of the sun, function as a rain protector and also multiply as Visual Barrier Sideways. The building thus functions as a selective environmental filter, enhancing the best components of the regional climate to address heating, cooling and ventilation needs of the structure.

生物气候设计

该项目遵循着一个综合的设计方法,以评估建筑节能并将其最大化。使用太阳能研究和模拟以产生日光照明、阴影分析、降雨模式数据和遮阳系统。这些研究,连同照明分析,对生成负荷计算、定型和根据大小将机械系统排列都有着至关重要的作用。为了优化制冷效果,建筑体和窗口的形状和大小都是根据计算机生成模拟被设计为最佳方式来捕获自然微风。垂直遮阳设备与水平遮阳设备结合,将炎烈的太阳光线隔离开来,这个设备同时可以作为遮雨设施并产生横向视觉遮挡。因此,建筑的功能就像一个选择性的环境过滤器,增强区域环境中的最好因素以满足结构中供暖、制冷和通风的需求。

Construction resources

The construction deals intensely on using green materials to integrate multiple sustainable features into the project. Using materials like AAC Blocks, PV Cells for solar lighting, Solar Reflective Tiles for high heat reflectance, Double Glazed Windows, Permeable Paving & Rain Water Harvesting for Ground Water Recharge that save significant energy and minimise the carbon footprint.

While the concept of the building focuses on green factors of design and use of sustainable materials, the aesthetic character of the bulding is far from being compromised. With individual residential units being approximately 10,000 sqft in size, the ecological elements are conscientiously woven together with the luxury requirements of the project that conclusively expresses a comteporary response which further establishes a contextual relationship and giving each residence the highest degree of originality.

建设资源

建设过程积极使用绿色材料，将多个可持续特色整合到该项目方案中来。使用材料类似于蒸压加气混凝土砌块、光伏电池作太阳能照明，太阳能反光瓷砖，用于高温反射、双层玻璃窗、透水铺装和雨水回收装置用于地下水填充，以上为该系统的几项措施，可节省大量能源，并将碳足迹最小化。

建筑的概念依然专注于设计的绿色要素和可持续材料的使用，而建筑的审美特征远不会因此有所妥协。随着单个住宅单位的面积接近10,000平方英尺，生态元素就被切实地与项目的奢华要求交织在一起，该项目最终呈现的是一种现代的响应，进一步确定了一个相关的关系，并赋予每一个住宅最高程度的创新性。

图书在版编目（CIP）数据

建筑是怎样炼成的：建筑创意与策略 / 香港建筑理工出版社主编．
-- 北京：中国林业出版社，2017.6

ISBN 978-7-5038-8929-5

Ⅰ．①建… Ⅱ．①香… Ⅲ．①建筑学 Ⅳ．① TU-0

中国版本图书馆 CIP 数据核字 (2017) 第 066807 号

建筑是怎样炼成的 建筑创意与策略
主　　编：香港建筑科学出版社

中国林业出版社
责任编辑：李　顺
出版咨询：(010) 83143569

出　版：中国林业出版社（100009 北京西城区德内大街刘海胡同 7 号）
网　站：http://lycb.forestry.gov.cn/
印　刷：深圳市汇亿丰印刷科技有限公司
发　行：中国林业出版社
电　话：(010) 83143500
版　次：2018 年 2 月第 1 版
印　次：2018 年 2 月第 1 次
开　本：889mm×1194mm　1 / 16
印　张：20
字　数：200 千字
定　价：368.00 元